CAMBRIDGE COUNTY GEOGRAPHIES

SCOTLAND

General Editor: W. Murison, M.A.

ORKNEY AND SHETLAND

Cambridge County Geographies

ORKNEY
AND
SHETLAND

by

J. G. F. MOODIE HEDDLE
and

T. MAINLAND, F.E.I.S.
Headmaster, Bressay Public School

With Maps, Diagrams, and Illustrations

CAMBRIDGE
AT THE UNIVERSITY PRESS
1920

CAMBRIDGE UNIVERSITY PRESS
Cambridge, New York, Melbourne, Madrid, Cape Town,
Singapore, São Paulo, Delhi, Mexico City

Cambridge University Press
The Edinburgh Building, Cambridge CB2 8RU, UK

Published in the United States of America by Cambridge University Press, New York

www.cambridge.org
Information on this title: www.cambridge.org/9781107646162

First published 1920
First paperback edition 2013

A catalogue record for this publication is available from the British Library

ISBN 978-1-107-64616-2 Paperback

CONTENTS

ORKNEY

CONTENTS

ILLUSTRATIONS

ORKNEY

ILLUSTRATIONS

The illustrations on pp. 3, 7, 9, 12, 27, 38, 46, 54, 62, 80, 81 are reproduced from photographs by Mr T. Kent, Kirkwall; that on p. 32 from a photograph by the Author; that on p. 40 by permission of Messrs J. Spence & Son, St Margaret's Hope; those on pp. 60 and 72 are from Tudor's *Orkney and Shetland*, by permission of Edward Stanford, Ltd., that on p. 69 by kind permission of Messrs William Peace & Son, Kirkwall; that on p. 71 from a photograph by Valentine & Sons, Ltd., that on p. 82 by kind permission of Dr Thomas Ross; and that on p. 88 by arrangement with Elliott & Fry, Ltd.

CONTENTS

SHETLAND

CONTENTS

ILLUSTRATIONS

SHETLAND

The illustrations on pp. 104, 112, 113, 118, 119, 162, are from photographs by Mr J. D. Ratter, Lerwick; those on pp. 117, 121, 123, 126, 132, 138, 140, 141, 143, 144, 147, 153, 154, 155, 159, 163, by Mr R. H. Ramsay, Lerwick; that on p. 137 by Valentine & Sons, Ltd., those on pp. 148, 149, 152, are from Tudor's *Orkney and Shetland* by permission of Edward Stanford, Ltd., and that on p. 158 by permission of Mr J. Nicolson of Glenmount, Lerwick.

ORKNEY

By J. G. F. MOODIE HEDDLE

PREFACE

I WISH to thank Captain Malcolm Laing of Crook for the photograph from Sir Henry Raeburn's portrait of Malcolm Laing, the historian; Andrew Wylie, Esquire, Provost of Stromness, for the portraits of Dr Rae and David Vedder; and J. A. Harvie-Brown, Esquire, Dunipace House, Stirlingshire, for the photograph of the Great Auk's resting-place.

J. G. F. M. H.

ORKNEY

1. County and Shire

The word *shire* is of Old English origin, and meant charge, administration. The Norman Conquest introduced an alternative designation, the word *county*—through Old French from Latin *comitatus*, which in mediaeval documents stands for shire. *County* denotes the district under a count, the king's *comes*, the equivalent of the older English term *earl*. This system of local administration entered Scotland as part of the Anglo-Norman influence that strongly affected our country after 1100.

The exceptional character of the historical nexus between the Orkney Islands and Scotland, makes it somewhat difficult to fix definitely the date at which Orkney can be fairly said to have first constituted a Scottish county. For a period of about one hundred and fifty years after the conditional and, to all intent, temporary cession of the Islands to Scotland in the year 1468, Scottish and Norse law overlapped each other to a large extent in Orkney. And although during that period the Scottish Crown both invested earls, and appointed sheriffs of Orkney, yet so long as Norse law subsisted in the Islands, as it did largely in practice and absolutely in theory until the year 1612, it is hardly possible to consider Orkney a Scottish county. The

A

relation of the Islands towards Scotland during this confused period of fiscal evolution bears more resemblance to that of the Isle of Man towards England at the present day. When, however, in the year 1612 an Act of the Scottish Privy Council applied the general law of Scotland to the Islands, although the proceeding was in defiance of the conditions of their cession, Orkney may be held to have at last entered into the full comity of Scottish civil life, and may thenceforth, without impropriety or cavil, be considered and spoken of as the County or Shire of Orkney.

The Latin name *Orcades* implies the islands adjacent to Cape Orcas, a promontory first mentioned by Diodorus Siculus about 57 B.C. as one of the northern extremities of Britain, and commonly held to be Dunnet Head.

The Norse name was *Orkneyar*, of which our *Orkney* is a curtailment. The name *orc* appears to have been applied by both Celtic and Teutonic races to some half-mythical sea-monster, which according to Ariosto, in *Orlando Furioso*, devoured men and women ; but the suggested connection between this animal and the name of the county appears a little far-fetched, although the large number of whales in the surrounding waters is quoted to support it.

2. General Characteristics and Natural Conditions

Orkney occupies the somewhat anomalous position of being a wholly insular shire whose economic interests are overwhelmingly agricultural. Most of the islands are flat or low ; and in several, such as Shapinsay, Stronsay, Sanday, and South Ronaldshay, the proportion of cultivated land exceeds 70 per cent. of their total areas. In the Mainland, however, there are large stretches of hill and moorland, while in Hoy and Walls

Rackwick, Hoy

the natural conditions of by far the greater portion of the island closely approximate to those of the Scottish Highlands. Rousay is the only other island which is to a large extent hilly ; but Westray and Eday have some hills, and Burray, Flotta, and several other islands considerable stretches of low-lying moor. The general rise of the land is from N.E. to S.W. A height of 334 feet is attained at the Ward Hill at the south end of Eday,

880 feet at the Ward Hill of Orphir, in the S.W. of the Mainland, 1420 at Cuilags, 1564 at Ward Hill, and 1309 at Knap of Trowieglen, the three highest points in Hoy and in the whole group. Exceptionally fine views are obtained from Wideford Hill (741 feet), near Kirkwall, and from the Ward Hill in Hoy, the varied panoramas of islands, sounds, and lakes perhaps gaining in grace of outline more than they lose in richness of detail from the woodless character of the country.

Taken in detail, and viewed from the low ground, however, the general aspect of much of the country is bleak, and only redeemed from baldness by the widely-spread evidence of a vigorous cultivation. Yet for reasons of a somewhat complex texture, involving meteorological conditions, historical and archaeological considerations, and a touch of all-round individuality, the Islands rarely fail to cast a spell upon the visitor. One might quote many distinguished writers to vouch for this fact, but an Orcadian poet has depicted the telling features of his native land, both physical and psychic, with unerring accuracy and skill.

> Land of the whirlpool, torrent, foam,
>   Where oceans meet in maddening shock ;
> The beetling cliff, the shelving holm,
>   The dark, insidious rock ;
> Land of the bleak, the treeless moor,
>   The sterile mountain, seared and riven ;
> The shapeless cairn, the ruined tower,
>   Scathed by the bolts of heaven ;
> The yawning gulf, the treacherous sand ;
> I love thee still, my native land !

Land of the dark, the Runic rhyme,
   The mystic ring, the cavern hoar,
The Scandinavian seer, sublime
   In legendary lore ;
Land of a thousand sea-kings' graves—
   Those tameless spirits of the past,
Fierce as their subject Arctic waves,
   Or hyperborean blast ;
Though polar billows round thee foam,
I love thee !—thou wert once my home.

With glowing heart and island lyre,
   Ah ! would some native bard arise
To sing, with all a poet's fire,
   Thy stern sublimities—
The roaring flood, the rushing stream,
   The promontory wild and bare,
The pyramid where sea-birds scream
   Aloft in middle air,
The Druid temple on the heath,
Old even beyond tradition's breath.

If we allow a little for the softer side of the picture, a side perhaps best typified by the fine old buildings of the little island capital, and the spell of the lightful midsummer night, which is no night, the lines of Vedder form a fair compendium of the natural conditions and general characteristics of the Islands to-day, although much of the " bleak and treeless moor " of the poet's youth has long since been converted into smiling fields of corn.

3. Size.   Situation.   Boundaries

The Orkney Islands extend between the parallels 58° 41′ and 59° 24′ of north latitude, and 2° 22′ and

3° 26′ of west longitude. They measure 56 miles from north-east to south-west, and 29 miles from east to west, and cover 240,476 acres or 375.5 square miles, exclusive of fresh water lochs. The group is bounded by the North Sea and the Pentland Firth on the south, the Atlantic on the west, Sumburgh Roost on the north, and the North Sea on the east. Our measurements take no account of the distant Sule Skerry, an islet of 35 acres lying 32½ miles north-west of Hoy Head, and inhabited only by lightkeepers and innumerable birds. The archipelago is naturally divided into three sections : the Mainland in the centre, the South Isles including all islands to the south, and the North Isles all to the north, of the Mainland. The Mainland—the Norse *Meginland*, or *Hrossey*, *i.e.* Horse Island—covers 190 square miles, and is 25 miles long from north-west to south-east, and 15 miles broad from east to west. It is divided into two unequal portions, the East Mainland and the West Mainland, by an isthmus less than two miles across, which connects Kirkwall Bay on its north sea-board with Scapa Flow, a large and picturesque inland sea, now well known as a naval base, which lies between its south coasts and the encircling South Isles. The name Pomona, stamped on the Mainland by George Buchanan's misapprehension of a Latin text, is never applied to the island by Orcadians ; and here be it also said that Hoy and Walls, the largest of the South Isles and the second in size of the whole group (13½ miles long by about 5½ miles broad, and covering 36,674 acres), although one island geographically, is colloquially

two. The principal of the other South Isles are South
Ronaldshay, 13,080 acres ; Burray, 2682 acres ; Flotta,
2661 acres ; Græmsay, 1151 acres ; and Fara, 840 acres.
Scapa Flow is connected with the Pentland Firth to
the south by Hoxa Sound, with the Atlantic to the
west by Hoy Sound, and with the North Sea to the east

The Home Fleet in Scapa Flow

by Holm Sound. The largest of the North Isles are
Sanday, 16,498 acres ; Westray, 13,096 acres ; Rousay,
11,937 acres ; Stronsay, 9839 acres ; Eday, 7371
acres ; Shapinsay, 7171 acres ; Papa Westray, 2403
acres ; North Ronaldshay, 2386 acres ; and Egilsay,
1636 acres. This section of the archipelago is itself
divided into two portions by the waterway formed by
the Stronsay and Westray Firths, which runs from

south-east to north-west through the islands, and offers
an alternative to the Pentland Firth or Sumburgh
Roost passages for vessels passing between the North
Sea and the Atlantic.

The whole archipelago includes some 67 islands,
besides a score or more of islets, but only 30 are inhabited,
and four or five of these are occupied solely by lighthouse attendants and their families.   Small uninhabited
islands many of which are used as pasturage, are known
as " holms."   The largest uninhabited island is the
Calf of Eday, of 590 acres, but some 10 of the inhabited
islands are of less area than this.

4. Streams and Lakes

The streams of Orkney are, of course, mere burns of
a few miles in length, draining the high ground, and
save for the cheap motive power which they offer to
farmers and millers, of interest only to anglers.   Nowhere
in Orkney are trees so much missed as along the burnsides,
and for that reason the Burn of Berriedale, a branch
of the larger Rackwick Burn, in Hoy, whose steep banks
are covered with poplar, birch, hazel, and mountain ash,
is a sort of Mecca to the aesthetic Orcadian.   The broad
estuary of the Loch of Stenness, which disembogues
into the Bay of Ireland, though of trifling length, is the
nearest approach to a river that the Islands can present.
It is known as the *Bush*.

From a spectacular point of view the many lakes of
Orkney go far to compensate the county for the absence

of rivers. So many are they that it is hardly possible to get out of sight of salt or fresh water in the Islands. The great twin lochs of Harray and Stenness, at once joined and separated by the Bridge of Brogar, give a strong dash of picturesqueness to the whole central part of the West Mainland ; and still more beautiful

Loch of Kirbuster, in Orphir

are the secluded and hill-surrounded Heldale Water and Hoglinns Water in Walls. Other lochs are Board-house, Swannay, Hundland, Isbister, Skaill, Banks, Sabiston, Clumly and Bosquoy in the West Mainland ; Kirbuster in Orphir, Tankerness in the East Mainland ; Muckle Water, Peerie (*i.e.* little) Water, and Wasbister in Rousay ; St Tredwell Loch in Papa Westray ; Saintear and Swartmill in Westray ; Muckle Water in

Stronsay ; and Bea, Longmay, and North Loch in
Sanday. The total area of Orkney lakes is about 20,000
acres. In most of them fishing is free, in others permission to fish is readily obtainable. The Loch Stenness
trout is, according to Gunther, a distinct species, but
this is a vexed question among the learned in such
matters. Heldale Water contains Norwegian char in
addition to trout.

5. Geology and Soil

The geological formation of the Orkney Islands is,
in its main features, of a very simple nature. If we
except a comparatively small strip of land running
northwestward from Stromness to Inganess in the West
Mainland, and a still smaller patch in the neighbouring
island of Graemsay, practically the whole county is
underlain by the Old Red Sandstone formation. The
two small areas above-mentioned—practically one,
save for the intervening sea—are occupied by older
crystalline rocks, consisting of fine-grained granite,
and micaceous schist at times running into foliated
granite, the whole traversed by veins of pink felsite.
These formations are flanked on either side by a narrow
band of conglomerate of the Middle Old Red Sandstone
period, composed of the rounded pebbles of the underlying schist and granite. Of the Old Red Sandstone
two divisions are found in the Islands—the Middle or
Orcadian, and the Upper Old Red Sandstone. There
are also in various parts but especially in Hoy and

Deerness, local outcrops of volcanic material. Orkney thus constitutes practically a continuation of the north-eastern Highlands of Scotland, and, as in Caithness, flagstone is by far the most widely distributed rock.

The Orcadian or Middle Old Red Sandstone is known to be a fresh water deposit, laid down by a large system of lakes which once extended over the north-eastern Highlands as far south as Inverness and Banff, and the same species of fossil fishes are found in the grey and black flagstones near Stromness as in the red sandstones on the south coasts of the Moray Firth. Leaving out of account the small area of crystalline rocks near Stromness, which are of age long anterior to any of the Old Red Sandstone deposits, the West Mainland parishes of Stromness, Sandwick, and Birsay contain the oldest rocks within the county, the Stromness flagstones with the fishes embedded in them being coeval with the Achanarras beds of Caithness. Next in point of age, come the flagstones of the East Mainland and the North Isles, corresponding in date and fossil remains to what are known to geologists as the Thurso Beds of Caithness. Following these in order of antiquity come the yellow and red sandstones and the dark red clays or marls which occur in the northern part of South Ronaldshay, and in Deerness, Shapinsay, and Eday. These are known as the John O' Groats Beds, from their occurrence in that part of Caithness also, where indeed they were first exhaustively investigated. These three members of the Middle or Orcadian Old Red Sandstone are supposed to have a combined thickness of not less than 7000 feet, though this pile of sediment represents only the upper half of the formation, the lower beds of Caithness apparently not being represented north of the Pentland Firth.

After the Middle Old Red Sandstone had stood for many ages as dry land, the eroding action of the atmo-

Berry Head, Hoy

sphere, rain, and streams gradually eliminated all inequalities and produced a nearly level surface, upon which by degrees a new lake was formed, occupying a large part of Orkney and Caithness. In this lake sand accumulated, which now forms the highest hills of Orkney. The yellow sandstones of the Hoy and Walls hills are deposits of this lake, and belong to the Upper Old Red Sandstone, which is not represented in the other islands of the archipelago, but reappears at Dunnet Head on the south side of the Pentland Firth.

Flagstone is a material that yields readily to the influence of the weather, and the subsoil formed is a rubble consisting of loose fragments of rock embedded in a brown clay formed by the softer and more weathered portion of the underlying rock. This gives a well-drained soil, and as the flagstones always contain lime, potash, and phosphates, and are frequently mixed with sand, the resultant soil may be generally described as a clayey loam of moderate fertility. Soil of this character covers a large part of the flagstone formation of Orkney, especially where the land is of moderate elevation, and surface accumulations of boulder-clay, or alluvium are at a minimum. At higher elevations on the other hand the character of the flagstone soils is frequently modified by the presence of peat. The soils of the sandstone formation, which occur in parts of South Ronaldshay and Hoy, and to a larger extent in Burray and Eday, are of inferior fertility, partly on account of their more porous nature. In Burray and Eday a large portion of the sandstone districts are in consequence left uncultivated, and utilised only as rough pasture. The same is true of the soils of the Upper Old Red Sandstone formation, which covers by far the larger portion of the parishes of Hoy and Walls, but most of this land is, in any case, above the altitude to which cultivation is usually carried in Orkney.

Of " drifts," or loose surface deposits, the boulder-clay is the only one that materially affects the agricultural quality of the soils of the Islands as a whole. The boulder-clay of Orkney, which on account of the particular direction—northwestwards over the Islands from the North Sea—taken by the ice at the period when the archipelago was subjected to glaciation, contains a considerable additional quantity of lime in the form of shells scraped up from the bed of the ocean, overlies the rock formations in the lower grounds over a large portion of the county. This deposit, which varies in thickness from less than a foot up to forty feet or more, has been calculated to occupy at least one-third of the area of the Islands, and it provides the most fertile soil that the county contains, particularly where the land has been improved by long cultivation and artificial drainage.

6. Natural History

The outstanding features of Orcadian zoology are naturally the very restricted number of land mammals as compared with that of the neighbouring mainland of Scotland, the relatively large number of cetaceans in the surrounding waters, and, above all, the richness of the avifauna, particularly in sea-birds, and autumn and winter visitants from more northerly climes.

Although the bones and antlers of the red deer have been found among the matter excavated from the sites of brochs and Picts' houses, and their shed horns at times turn up among the peat-mosses of the Mainland, that king of British Cervidæ was unknown to Orkney during historic times until about the year 1860, when two young hinds and a young stag were introduced into Walls. There they throve perfectly, and had increased to thirteen or fourteen by 1870-72, when the proprietor of the island found it necessary to kill them off, on account of the damage which they were doing to crops. Tusks of the wild boar have been found at Skaill in the West Mainland, but the wolf, fox, badger, and in fact practically all of the larger land mammals known to Britain during historic times, or still found there to-day, have been totally unknown in Orkney during the same period. The otter is a notable exception, as it is very abundant in most of the islands, the great extent of seaboard giving it special facilities for concealment and avoidance of capture. An Orcadian proprietor who died a few years ago has recorded that in his young days he often had as many as thirty otter skins in his possession at one time.

The common hare appears to have been introduced into Orkney by a Mr Moodie of Melsetter, early in the eighteenth century, but both that attempt at acclimatisation and one by Malcolm Laing, the historian, in 1818, proved comparative failures. Better success, however, attended the efforts of Mr Samuel Laing and Mr Baikie of Tankerness about 1830, and at the present day, hares are found in the Mainland, Rousay, Eday, Shapinsay, Hoy, and South Ronaldshay. The white hare occurred in Hoy at an early date, as recorded by Jo Ben, a resident in Orkney, in his *Descriptio Insularum Orchadiarum*, in 1529 :—" Albi lepores hic sunt, et capiuntur canibus."

It died out, however, and has only recently been re-introduced. The rabbit is common throughout the Islands. Of other and less desirable rodents, the black rat was once general in South Ronaldshay, and probably still occurs there. The brown rats, common in most parts of Orkney, have been known at times to forsake certain islands altogether, taking to the sea in a body in search of a new home. The field mouse, and the house mouse are universal. The common field vole is plentiful in most of the islands, and there is a doubtful record of the water vole from Hoy. *Microtus orcadensis*, or the Orkney vole, discovered in 1904, is a highly interesting species peculiar to Orkney and certain parts of Shetland. The common shrew has been found in Walls and Orphir, and the water shrew in Walls. Bats are rare in Orkney, but occurrences of *vesperugo pipistrellus* have been recorded from Walls, Sanday, and Kirkwall, of *vespertilio murinus* from Walls ; while there is an interesting but doubtful record of a specimen of *vesperugo noctula* having been captured in South Ronaldshay.

Appearances of the walrus in Orkney waters have been recorded from Eday, Hoy Sound, and Walls at various dates from 1825 to 1864 ; and Orkney seals include *phoca vitulina*, which breeds on several islands and skerries, *phoca groenlandica*, and the grey seal. The occurrence of the hooded seal is doubtful. Among Cetacea, the Greenland and sperm whales are rare visitors, the hump-backed whale, still rarer ; but the common rorquall, Sibbald's rorquall, the lesser rorquall, the beaked whale, the grampus, the common porpoise, and the white-sided dolphin are all fairly common. The bottlenose is, however, *the* Orkney whale, occurring at times in schools of 500 in number. The bottlenosed dolphin and the white-beaked dolphin are also on record.

The ornithology of Orkney comprises about 235 species, and owing to the special physical characteristics of the Islands bird-life forms a more conspicuous feature of landscape and sea than it does perhaps anywhere else in the British Islands. In a district where travel is more usual on sea than on land, and where the lakes, the fields, the hills, and the moors are unshrouded by woods, not only are aquatic birds a more constant object of the view than in districts otherwise conditioned, but the commoner land birds also are more frequently and readily observed.

Of the Falconidæ 17 species have been killed or observed in the Islands, being practically all of this family known to Britain, except the orange-legged falcon and the bee hawk. The golden eagle and the white-tailed eagle, however, both of which formerly bred in Hoy, are now only occasional visitants. The peregrine falcon is still fairly common, and in old days the King's falconer procured them from the Islands for sporting purposes. Of the Strigidæ, the long-eared owl, the short-eared owl, the snowy owl, the tawny owl, Tengmalm's owl, and the eagle owl have all been observed ; but recorded occurrences of the barn owl and the little owl are of doubtful authenticity. Of the

B

order Anseres, of which some 32 species have been observed in the Islands, the rarest locally are perhaps the greylag goose, the pink-footed goose, the Canada goose, the gadwall, the shoveller, the Garganey teal,

The Great Auk
(*Alea Impennis*)

the king eider, the harlequin duck, the common scoter, the surf scoter, Bewick's swan, and the goosander. The common eider duck is plentiful. Regular winter visitants, but not unknown at other seasons, are the

long-tailed duck, the velvet scoter, and the smew.
Other visitants are the bernacle goose, the brent goose,

Crannie in which last Great Auk lived

the white-fronted goose, and the hooper, the last two
in particular being common frequenters of the larger

lochs, such as Stenness, Harray, and Boardhouse at this season. Of some 18 species of the Laridæ found in Orkney, the rarest are perhaps the Iceland gull, the glaucous gull, the common skua, the pomatorhine skua, and Richardson's skua, the last-named, however, breeding in Walls. Of the Colymbidæ, the red-throated diver breeds in Walls, while the black-throated and great northern divers are winter visitants, both suspected to have occasionally stayed to breed. Of the Alcidae, the razor-bill, common guillemot, black guillemot, puffin, and little auk are usual. In the crevice of a cliff in Papa Westray lived the last Orkney great auk, *alca impennis*, shot in 1813, and now in the Natural History Museum, South Kensington, an interesting specimen of a bird probably now everywhere extinct. Of the Scolopacidae about 22 species are on record. Yarrell is perhaps in error when he mentions the avocet as having been found in the Islands, but the red-necked phalarope was first recorded as a British species from Stronsay in 1769. The gray phalarope is rare. The woodcock comes in winter, and has bred in Rousay. The common snipe is plentiful, and the jack snipe and double snipe come in autumn. The little stint, purple sandpiper, sanderling, knot, ruff, bartailed godwit, black-tailed godwit, spotted redshank, and the greenshank are all found, but some of these are rare. Of the Charadriidae, the golden, grey, and ringed plovers, the lapwing, dotterel, and turnstone are common, while the eastern golden plover has been found. Of the order Tubinares, the stormy petrel and the Manx shearwater breed, and the fulmar petrel has become a common visitant of recent years. The Rallidae are represented by the land-rail, water-rail, spotted crake, moor-hen, and common coot, and the Gruidae by the common and demoiselle cranes. The common heron alone is usual among the Herodii, although the bittern, little bittern, white stork, spoonbill, and glossy ibis have all been found. Of the Podicipitidae, the Sclavonian, great-crested, eared, and little grebes are known, the last-named, however, being the only nester. Of the Pelicanidae, the solan goose breeds on the distant Stack, near Sule Skerry, and the shag and cormorant are common. Of the Columbidae the rock-dove is common, the ring-dove occasionally breeds in plantations, and the stock-dove and turtle-dove are seen at times. Of the order Passeres, the usual nesters include the song-thrush, blackbird, redbreast, wren, pied wagtail, rock pipit, linnet, twite, greenfinch, yellow bunting, skylark, and common starling, the last a bird perhaps more frequent in the Islands than anywhere else. Scarcer breeders are the missel-thrush, stonechat, ring-ouzel, golden-crested wren, sedge warbler, grey wagtail, yellow wagtail, hedge-sparrow, meadow pipit, pied flycatcher, swallow, sand-martin, chaffinch, lesser red-pole, and reed bunting. Common winter visitants are the redwing, fieldfare, and snow bunting, while rarer or only occasional comers are the dipper, redstart, black-cap, chiffchaff, fire-crested wren, willow wren, great titmouse, blue titmouse, common creeper, grey-headed wagtail, tree pipit, great grey shrike, red-backed shrike.

waxwing, spotted flycatcher, rose pastor, goldfinch, brambling, mealy redpole, common bullfinch, common crossbill, and wood lark. Orkney Corvidae include the grey crow, the rook, the jackdaw (South Ronaldshay only), and the raven as breeding species, while the magpie and nutcracker are rare visitants. Of the Picidae, the great spotted woodpecker is an irregular autumn and winter visitant, while the lesser spotted woodpecker, the green woodpecker, and the wryneck are seen at times. Of the order Coccyges, the cuckoo is fairly common, while the roller, hoopoe, and common kingfisher have been found. Of the order Macrochires the common swift and the common nightjar are occasionally seen.

Of game and other sport-yielding birds, the red grouse breeds in the Mainland, Rousay, Eday, Hoy, Walls, Flotta and Fara. Grouse disease is unknown in Orkney, and the birds of Walls and Rousay are the heaviest in Scotland. Various attempts to acclimatise the black grouse, partridge, red-legged partridge, and pheasant have all practically failed. The ptarmigan bred in Hoy until 1831. Pallas's sand grouse at times visit the Islands in considerable numbers, and are surmised to have bred in several islands. The quail comes in much the same way, if in fewer numbers, and has nested, though rarely.

The plant-life of the Islands, however interesting to the scientifically-equipped botanist, presents no such happy hunting-ground to the unsophisticated lover of wild nature as does their bird-life. The practical

non-existence of woods conspires with the cool summer
and high winds of the country to restrict both the
number and the distribution of its flora. Ferns in
particular are of circumscribed distribution, a loss to
the beauty of the country-side only less conspicuous
than that caused by the absence of woods ; while
several other popular and showy plants, such as the
wild rose, the foxglove, gorse, and broom are of only
too limited a range. Some 20 species or varieties of
ferns are known or reported, of which *ophioglossum
vulgatum, var. ambiguum*, was for years unknown out
of the Islands. *Zannichellia polycarpa*, a pond-weed,
was for some time known as a British plant only from
the Loch of Kirbuster in Orphir ; and *Carex fulva*, a
sedge, was at one period peculiar to the same parish.
More interesting, however, is the recent discovery by
Mr Magnus Spence, who has lately published the first
complete *Flora orcadensis*, of a plant which Mr C. E.
Moss, D.Sc., of Cambridge, considers to be either a
new variety of the dainty *Primula scotica*, or *Primula
stricta*, a species hitherto unknown to the flora of the
British Isles. The common variety of *Primula scotica*
is fairly abundant in many of the islands. Hoy is the
most interesting of the islands from a botanical point
of view, as it contains a variety of plants unknown to
the others. Perhaps the most interesting of these is
Loiseleuria procumbens, the trailing azalea, which
makes a beautiful show in its season on several spots
among the higher hills. This island also contains in
several of its more sheltered glens practically the only

indigenous trees that Orkney can boast of, consisting of somewhat stunted specimens of hazel, birch, mountain ash, quaking poplar, and honeysuckle. Before the days of the Baltic timber trade the dying Orcadian must have been gravely concerned over the disposition of the family porridge-stick, or " pot-tree," as it was locally styled. Even to-day, with some plantations around certain mansion-houses, it is doubtful whether all the trees in the county, indigenous and introduced, would cover a sixty-acre field. We subjoin a list of a few of the rarer Orkney plants, with some of their localities.

Thalictrum Alpinum	Hills of Hoy, Orphir, Rousay.
,, Dunense	Links, in Walls, Deerness, Sanday.
Ranunculus Sceleratus	Stromness.
Nasturtium Palustre	North Ronaldshay.
Sisymbrium Thalianum	Kirkwall, Hoy.
,, Officinale	Hoy.
Draba Incana	Hoy Hill ; Fitty Hill, Westray.
Silene Acaulis	,, ,, ,,
Spergularia Marginata	The Ayre, Walls ; Vaval, Westray.
Geranium Robertianum	Carness, St Ola.
Fragaria Vesca	Rousay.
Rubus Fissus	Hoy.
Dryas Ocopetala	Hoy Hill ; Kame of Hoy.
Rosa Glauca, *var.* crepiniana	Stromness.
Circæa Alpina	Hoy, Orphir, Evie.
Sedum Acre	Links, Hoxa, S. Ronaldshay.
Saxifraga Oppositifolia	Hoy Hills.
,, Stellaris	Rackwick, Hoy ; Kame of Hoy.
Pimpinella Saxifraga	St Ola.
Sium Erectum	Holm, Sanday.

Hedera Helix . . .	Berriedale, Hoy ; Berstane, St Ola.
Cornus Suecica . . .	Kame of Hoy.
Gallium Mollugo, *var.* Bakeri	Deerness, Westray.
Hieracium Orcadense . .	Cliffs in Hoy.
,, Scoticum .	Cliffs in Orphir.
,, Strictum .	Pegal Bay, Walls.
,, Auratum .	Cliffs, Pegal Bay, etc.
Lobellia Dortmanna .	Walls, Rousay.
Jasione Montana . .	Eday, N. Ronaldshay.
Arctostophylos Uva-Ursi .	Hills, Hoy and Walls.
Pyrola Rotundifolia .	Hoy, Rousay.
Vaccinium Vitis-Idæa .	Walls, Hoy, Orphir, Rousay.
,, Uliginosum .	Walls, Hoy, Birsay.
Gentiana Baltica . .	North Ronaldshay, Birsay.
Ajuga Pyramidalis .	Berriedale and Rackwick, Hoy.
Myosotis Palustris .	Orphir, St Andrews.
Oxyria Reniformis .	Hoy.
Myrica Gale . . .	Birsay.
Salix Nigricans . .	Orphir.
Juniperus Nana . .	Hoy.
Typha Latifolia .	Loch of Aikerness, Evie.
Sparganium Affine .	Mainland, Hoy, Rousay.
Ruppia Spiralis . .	Loch of Stenness.
,, Rostellata, *var.* Nana .	Oyce of Firth.
Goodyera Repens .	Stromness, Harray.
Scirpus Tabernæmontani .	St Ola, Holm.
Blysmus Rufus . .	Orphir, St Andrews, Westray.
Carex Muricata . .	Firth.

The total number of plants found in the Islands, not counting varieties, is about 560, a number slightly in excess of that of Shetland and slightly fewer than that of Caithness, to the floras of which counties that of Orkney closely assimilates.

7. The Coast

In a district where, despite the general existence of good roads, the shortest cut to church, post office, smithy, or mill is often by crossing a sound or skirting the shore in a yawl, the coastline spells something more than a mere alternation of cliffs and sandy beaches, diversified by the occasional appearance of a lighthouse or a harbour. Such things of course the shores of Orkney exhibit in no common measure, but to show how far they are from exhausting the coastal features of Orcadian life and scenery, it is only necessary to say that of twenty-one civil parishes in the county all but one (Harray) possess miles of sea-board, and that of the centres of population, only some two or three hamlets are inland. No spot in the Mainland is above five miles from the coast, no point in any other island more than three miles.

The general configuration of the coasts may be best studied on the map, but as not every inlet of the sea forms a good natural anchorage, we here indicate some that do so. Others, and some of these the most important, are mentioned in the final section. Widewall Bay on the W. side of South Ronaldshay, Panhope in Flotta, and Echnaloch on the N.W. of Burray are, after the far-famed Longhope in Walls, the best anchorages in the South Isles. The Bay of Ireland, known to mariners as Cairston Roads, is on the south coast of Mainland, a little to the eastward of Stromness. Inganess Bay and Deer Sound are in the N.E. of Mainland ; Veantrow Bay on the N. side of Shapinsay ; St Catherine's Bay, Mill Bay, and Holland Bay in Stronsay ; and Otterswick in Sanday. There are deepwater piers—as indispensable adjuncts of traffic in

The Old Man of Hoy

(The tallest "Stack" in British Isles, 450 ft. high)

Orkney as railway stations are elsewhere—at Longhope, St Margaret's Hope, and Burray in the South Isles ; at Stromness, Swanbister Bay, Scapa Bay, and Holm on the S. coast of Mainland ; at Kirkwall and Finstown on the N. coast of Mainland ; and in Shapinsay, Stronsay, Eday, Rousay, Sanday Westray, Egilsay, and North Ronaldshay in the North Isles.

Lying as they do athwart a main trade route from the ports of northern and western Europe to America and the western ports of Britain, the rock-girded and wind-swept shores of Orkney are especially well lighted. Indeed it is probable that from the summit of the Ward Hill of Hoy on a clear night more lighthouses can be discerned than from any other point in Britain. Stromness is an important centre of operations for the Scottish Lighthouse Board, lying half-way between the east and west coasts of Scotland, and having the many lighthouses of Shetland to the north.

The coasts and sounds of Orkney are studded with innumerable skerries and sunken reefs. Few of these call for any special notice, but the reader should know that in a few cases—Auskerry, the larger Pentland Skerry, Sule Skerry—the name skerry is locally applied to soil-covered islets of considerable area. There are no raised beaches in Orkney, all oceanological and geological data going to show that the islands were once united, and the coastline in consequence at a lower level than it occupies to-day. Traces of submerged forests are to be found at Widewall Bay, in South Ronaldshay, and a few other localities. Coast erosion in the Islands is too slight a factor to have any practical significance.

Apart from the fine natural harbours, the outstanding physical feature of the Orkney coasts is the gigantic cliff scenery of the Atlantic sea-board, particularly that of Hoy and Walls, which both for loftiness and splendour of colouring stands unrivalled in Britain.

For a distance of about two miles from the Kaim of Hoy southwards to the Sow the average height of the cliffs is above 1000 feet, the huge rampant culminating about midway between these two points in St John's Head, also called Braeborough, which attains a height of 1140 feet. About a mile southward of the Sow stands the famous Old Man of Hoy (450 feet), tallest of British " stacks "—

> A giant that hath warred with heaven,
> Whose ruined scalp seems thunder-riven.

Of many other fine cliffs in this island we have space to mention only the Berry, in Walls, a sheer precipice of 600 feet in height, which forms, so to speak, one of the jaws of the Pentland Firth, the other jaw being Dunnet Head on the Caithness side. For beauty of colouring and indeed of outline, the Berry excels any cliff in this wonderful coastline, which the late Dr Guthrie described as, after Niagara and the Alps, the most sublime sight in the world. There is much fine rock scenery elsewhere in the Islands, particularly in the West Mainland, Rousay, Eday, and South Ronaldshay.

The tideways in many of the sounds which separate the various islands are remarkable for their turbulence and velocity, a speed of 8 knots an hour being reached in Hoy Sound, and 12 knots in the Pentland Firth. In the Pentland Firth also are two whirlpools, the Wells of Swona and the Swelkie of Stroma, the latter as famous in fable as the by no means more formidable

Maelström. The total coastline of the Islands extends to between 500 and 600 miles.

8. Weather and Climate

Insular climates are almost invariably milder than those of continents, or even those of the inland regions of large islands, in the same latitudes, and the climate of Orkney is no exception to this rule. Like so many other things Orcadian, the climate is conditioned by the proximity of the sea, and in this case by a sea whose waters are considerably warmer than their latitude might lead one to suspect. The warm surface drift of the North Atlantic is of itself sufficient to explain the relatively mild winter of Orkney, and the presence of the widest portion of the North Sea on the eastern side of the Islands has also a modifying influence. It must also be borne in mind that in cool climates rain brings heat. Our westerly and south-westerly winds, passing over sun-bathed seas, collect in their courses the vapour of warmer climates, and when this vapour, coming into contact with the cooler air of more northerly latitudes, is again condensed into water, a certain amount of the heat thus collected is set free, and raises the temperature of the air, of the rain itself, and of the land on which it falls.

TABULAR STATEMENT OF ORKNEY WEATHER.

	Mean Temperature. Years 1871-1905.	Mean Hours Sunshine. 1880-1907.	Mean Rainfall. 1841-1907.
January .	. 39.0°	29.7	3.72″
February .	. 38.5	55.5	3.05
March .	. 39.3	101.1	2.82
April . .	. 42.4	154.1	1.99
May . .	. 46.4	178.5	1.81
June . .	. 51.3	160.9	1.97
July . .	. 54.2	141.3	2.57
August .	. 54.0	121.8	3.01
September .	. 51.5	108.8	3.09
October .	. 46.4	75.5	4.43
November .	. 42.4	36.5	3.97
December .	. 39.9	20.8	4.21
Mean .	. 45.4	Total 1184.5	Total 36.65

It will thus be seen that the mean annual temperature of Orkney is 45.4°, which compares with 46.3° at Aberdeen and at Alnwick in Northumberland, and with 49.4° at Kew Observatory. The total range of temperature is only about 16°, as against 20° at Thurso, just across the Pentland Firth, 22° at Leith, and 25° at London. In this respect the Islands resemble the S.W. coast of England and the W. coast of Ireland. The lowest temperature recorded in Orkney in the eighty-seven years during which meteorological observations have been made was 8°, which occurred on 18th January 1881, the highest was 76°, on 16th July 1876. The temperature of the ocean varies only about 13° during the year, from 41.6° in February, to 54.5° in August. The mean annual rainfall of about 37 inches compares with over 80 inches

Old Melsetter House, in Walls
(In a specially mild winter climate)

in many parts of the West Highlands, and with 23 inches at Cromarty, the driest station in Scotland. The wettest months are October, November, and December, during which the Islands receive from one-third to one-half of their annual rainfall, the driest months are April, May, and June, which together receive only one-eighth of the total fall. Thus Orkney is practically never troubled with excessive rainfall, and serious droughts are equally unknown. The mean annual sunshine of 1185 hours compares favourably with 1164 hours at Edinburgh. London enjoys 1260 hours, Hastings 1780. Orkney's brightest month is May, with an average of 178 hours of sunshine, the gloomiest is December with 20.6 hours.

Apart from the fact that Orkney enjoys the mildest winter of any Scottish county, the chief difference between the weather of the archipelago and that of Scotland in general is perhaps the greater prevalence of high winds in the Islands, which owing to the general lowness of the land receive the full force of the North Atlantic gales, and which moreover lie in the most common track of the Atlantic cyclones, a circumstance which leads to great variability of wind and weather. Orkney has record of only one hurricane, on 17th November, 1893, with a velocity of 96 miles. Several winter gales of over 80 miles have been recorded, and one summer gale of 75 miles in the year 1890. During the fifteen years 1890-1904, 300 gales were recorded in Orkney, practically the same as at Fleetwood in Lancashire, while Alnwick experienced only 157, and Valentia on the west coast of Ireland only 130. Atlantic cyclones are the dominating factor in Orkney weather during the greater portion of the year, producing gales of greater or smaller magnitude, and being almost invariably accompanied by rain, with sudden changes both in the direction and the force of the wind. In the spring season, however, anticyclones frequently cause

c

spells of dry, cold weather, with fairly steady winds from the eastward or northward.

Winter in Orkney is in general a steady series of high winds, heavy rains, and ever varying storms, with much less frequent falls of snow, and fewer severe or continuous frosts than elsewhere in Scotland. Under the shelter of garden walls we have seen strawberry plants in blossom at Christmas and roses in January, while chance primroses may be found in sheltered nooks in any month of the year. The spring is cold and late, but the prevailing winds from N.W., N.E., or E. have not the piercing coldness so often felt in the spring winds along the east coast of Scotland. The summer is short, but remarkable for rapidity of growth. Fogs are fairly common during summer and early autumn, and come on and disperse with exceptional suddenness. Thunder in Orkney occurs mostly in winter, during high winds and continuous falls of rain or snow. The heaviest rains and the most prevalent and strongest winds are from the S.W. and S.E.

MEANS OF OBSERVATION IN ORKNEY FOR THIRTY-
THREE YEARS—1873–1905

Rainy days.	Snowy days.	Days on which Hail fell.	Thunder storms.	Clear sky.	Overcast.	Gales.
219	31	14	6	31	156	79

9. The People—Race, Language, Population

It is unsafe to dogmatise on the early races of Orkney ; but from the undoubted community in blood, speech, and culture with other northern counties of Scotland during the Celtic period, we may fairly conjecture that the Islands must have similarly shared in whatever pre-Celtic population—Iberian or other—these regions as a whole possessed.

Into the vexed question whether any remnant of Celtic population survived the Norse settlement of the Islands in the ninth century we cannot enter here. It is certain that for centuries after that era Norse speech, law, and custom were as universal and supreme in Orkney as ever Anglo-Saxon speech and institutions were in Kent or Norfolk. The place-names of the Islands are, save for a late Scottish and English element, entirely Norse.

The Scottish immigration into Orkney, which commenced about 1230, came for centuries almost exclusively from the Lowlands—the Lothians, Fife, Forfarshire, and those parts of Stirlingshire, Perthshire, Aberdeenshire, Banffshire, and Moray which lie outside the " Highland Line," being the chief areas drawn from. A later and much slighter strain of immigration from Caithness, Sutherland, and Ross, which affected the South Isles more especially, was itself quite as much of Norse as of Celtic ancestry. The people of Orkney must therefore be put down as in the main an amalgam of Norse and Lowland Scots.

There never was any very rigid line of division between these two races in the Islands. So much is it the case that, while the Norse were being Scotticised in speech and custom, the incomers were at the same time being Orcadianised in sentiment, that the opprobrious epithet of "Ferry-loupers," hurled by native Orcadians at successive generations of Scots intruders, is itself of Scottish origin. The *genius loci* was a very potent spirit, and the Scoto-Orcadian was often prouder of being an Orcadian than of being a Scot.

The humblest Orcadians have for centuries past spoken English more correctly and naturally than was at all common among the Scottish lower orders before the advent of board school education, a circumstance largely due to the fact that a great portion of the population exchanged Norse speech for Scots about the period—the later sixteenth and early seventeenth centuries—when educated Scots were themselves adopting English. Add to this the constant presence of passing English vessels and the presence of Cromwell's soldiers. At the same time a modified Scots dialect is commonly spoken by the less educated classes, but even in this the admixture of pure English is pronounced. A few Norse words, mostly nouns, survive imbedded in the local speech, whether English or Scots. Norse speech lingered in Harray until about 1750.

The population of Orkney, which numbered 24,445 in the year 1801, increased almost uninterruptedly to a maximum of 32,395 in 1861. Every subsequent census, except that of 1881, has shown a decrease, and the 9.8 per cent. rate of intercensal decline recorded in 1911 was the heaviest revealed by the census of that year for any Scottish county. The population in that year was 25,897, being 69 to the square mile.

Under modern conditions Orkneymen have resumed the roving instincts of their Norse ancestors, and the variety of capacities under which the sons and daughters of the Islands live in the far corners of the earth is astonishing. The ostensible local causes of this move-ment are in reality of secondary importance. The real causes are improved education, improved communi-cations, and the grit to take advantage of them. Can any Scottish or English county show the equivalent of *The Orkney and Shetland American*, a little news-paper published for years in Chicago ? " A Shields Shetlander," too, is a current descriptive tag which might well be supplemented by " A Leith Orcadian."

10. Agriculture

Farming is the very life of Orkney, giving full or partial employment to no less than 6400 of the popu-lation. The great era of agriculture in the Islands followed, and was partly the consequence of the failure of the local kelp industry in the second quarter of the nineteenth century. The area under crop and permanent pasture rose from about 30,000 acres in 1855 to 86,949 acres in 1870. It is now 107,941 acres, while in addition at least 52,941 acres of heath and mountain land are utilised for grazing. The chief crops are oats, 33,153 acres; turnips, 13,877 acres; and hay, 9425 acres. Stock rearing is the cornerstone of Orcadian farming.

Short-horns and polled Angus are the favoured breeds of cattle, and many thousand head from the Islands

Harvesting at Stenness

pass through the Aberdeen auction marts every year. The finest of the beef—and prime Orkney beef is second to none—finally reaches Smithfield. Cheviots and Cheviot-Leicester crosses are the common sheep, the small native breed, of Norwegian origin, being now confined to North Ronaldshay. The old Orkney horse, itself probably a hybrid of half Norwegian and half Scottish extraction, has for several generations past been crossed with Clydesdale blood, and the resultant is a small-sized but very sturdy and serviceable animal. Oxen are still used to a small extent as draught animals. The export of eggs and poultry is a great and growing Orcadian industry, the annual output from the Islands being at least £60,000 in value, a larger figure be it noted than the purely agricultural rental of the county. The fattening of geese for the Christmas market is a special feature of Orkney poultry-farming, the birds being largely brought from Shetland at the end of harvest and put on the stubble. The open winter is a valuable consideration to poultry-keepers in the Islands, increasing the amount of natural food which the birds are able to pick up, and extending the period of laying.

Large quantities of sea-weed are available as manure in practically every part of the Islands, and marl in some localities. The chief disadvantages under which agriculture labours in Orkney are distance from the markets, and occasional damage to grain crops from sea-gust. Agricultural co-operative societies, however, which have obtained a firm footing in the Islands, are doing a great deal to counter-act the effects of the first-mentioned drawback.

11. Industries and Manufactures

The manufacture of kelp was introduced into the Islands in 1722, and by 1826 the annual export amounted to 3500 tons, valued at £24,500. The abolition of the duty on barilla, which is largely used in the manufacture of glass, destroyed this industry for a time ; but since about 1880 there has been a considerable revival in the North Isles, the yearly export having again reached about 1500 tons. Orkney kelp is considered of the finest quality.

The making of linen yarn and cloth, introduced in 1747,

Orkney Yawl Boats

was successfully carried on for many years, and flax was locally grown. This industry received a severe check during the Great War (1793-1815), and gradually disappeared.

The manufacture of straw-plait for bonnets and hats was begun about 1800, and fifteen years later the yearly export was valued at £20,000, from 6000 to 7000 women being employed in the industry. The material used was at first split ripened wheat straw ; later, however, unripened, unsplit, boiled and bleached rye-straw was substituted. The reduction of the import duty on straw-plait finally destroyed this interesting home industry, of which Kirkwall and Stromness were the chief centres.

The present-day industries of Orkney are unimportant. No minerals are worked in the county, although flagstone is quarried at Clestrain in Orphir, and red sandstone of fine quality at Fersness in Eday. The numerous sailing boats used in the Islands are mostly of home construction, the broad-beamed, shallow-draught, and comparatively light Orkney yawl being a type specially designed to suit local conditions of weather and tide. The making of the well-known Orkney straw-backed chairs is restricted by a very limited demand, and the specimens made for sale are somewhat more elaborate than those used in the cottages. A small quantity of home-spun tweed is made in the Islands, and a certain amount of rough knitting—stockings, mittens, and other articles used by the seafaring classes—is done in some districts. Fish-curing is carried on at Kirkwall, and to a small extent, as a home industry, in country districts. There are distilleries at Stromness, Scapa, and Highland Park, near Kirkwall, the output at the last-named being large, and in high esteem among whisky-blenders.

12. Fisheries and Fishing Station

Of recent years Whitehall in Stronsay has become one of the great centres of the summer herring fishing, with an annual catch of from 80,000 to 90,000 crans, a total exceeded in Scotland only at the ports of Lerwick, Fraserburgh and Peterhead. As at many other places where this great industry is carried on, however, the boats, the capital, and the personnel come almost entirely from outside. There are smaller stations of this fishery at Kirkwall, Sanday, Stromness, Holm, and Burray, at the last-named of which alone the boats and crews are local.

The white fishing is carried on in a desultory fashion in Orkney waters by some 350 fishermen, who use small locally-made yawls, but the annual catch is not important compared with that of other fishery districts. Haddocks, cod, and saithe are the commonest fish. Saithe simply swarm, but are caught chiefly for household consumption. Some 250 other men who style themselves " crofter fishermen " in the census returns, are in reality small farmers who do an occasional day's fishing, mainly for the pot. Lobster fishing is the one branch of the industry which the " crofter-fishermen " does follow with any persistence, and lobsters and other shell-fish, mainly whelks, to the value of from £6000 to £7000, are exported from the Islands annually The whelks are gathered mainly by women. There are 338 fishing boats in Orkney, of an aggregate burden of 2154 tons, and of a value, including fishing gear, of £16,095.

Sea-trout are plentiful along much of the Orkney coast, especially near estuaries ; but although surreptitious netting is intermittently done by unauthorised persons, this fishing is not, as with proper care it might be, on any sound commercial footing. Sea-trout run to a large size in the Islands, fish of from 8 to 10 lbs. being not uncommon. Walls, Hoy, the Bay of Ireland, Holm, and Rousay are the best localities. The net season is from 24th February to 10th September, the rod season from the same opening date to 31st October.

Longhope and the Bay of Firth were of old famous for oysters, and at the latter place a praiseworthy effort was recently made to restore the fishery.

13. History of the County

Our knowledge of the Orkney Islands before the Norse settlement in the latter part of the ninth century is of a slight and fragmentary character. In particular, what Latin writers say gives no sure information, the references in poets like Juvenal and Claudian being manifestly for literary ornament.

The earliest writer of British race to throw any light on the Islands is Adamnan, who mentions that in the sixth century Cormac, a cleric of Iona, with certain companions, visited the Orkneys, and adds that the contemporary Pictish ruler of the Islands was a hostage in the hands of Brude Mac Meilcon, King of the Northern Picts. Whatever degree of power this Pictish king may have exercised over the Islands, we learn from the *Annals of Ulster* that in the year 580 they were invaded by Aidan, King of the Dalriadic Scots, and as the next mention of the Orkneys in the native chronicles is the record of their devastation by the Pictish King Brude Mac Bile in the year 682, it is perhaps a fair inference that Dalriadic influence had predominated there during the intervening century. That the Islands were christianised about this period by clerics of the Columban or Irish Church, is a point too firmly established by archæological, topographical, and other data to require any insistence on here. Many pre-Norse church dedications to St Columba, St Ninian, and other Celtic saints, tell their own story.

Little is known of the state of the Islands during two centuries preceding the date of the Norse settlement *c*. 872 A.D., but from that era until the year 1222 Orkney possesses in the *Orkneyinga Saga* a record of the highest value. The *Saga* states that the Islands were settled by the Norsemen in the days of Harald the Fair-haired (Harfagri), but had previously been a base for Vikings. Harald Harfagri had about the year 870 made himself sole King of Norway, and in so doing had incurred the odium of a large section of the Odallers, or landowners, many of whom in consequence emigrated to Iceland, Shetland, Orkney, the Hebrides, and the coasts of Ireland. The settlers in Orkney, Shetland, and the Western Islands took to piracy, and so inflicted the coasts of old Norway, that in 872 Harald followed up the fugitives, conquered all the islands of the Scottish seas, and placed his partisan Rognvald, Earl of Moeri, as hereditary Jarl over Orkney and Shetland. This nobleman, however, preferring to live in Norway, gifted his western jarldom to his brother Sigurd, who is commonly considered the first, as he proved one of the greatest, of the long line of Orcadian Jarls. Sigurd speedily spread his power over northern Scotland as far south as Moray, and from his time until the close of the thirteenth century the Orkney Jarls had the controlling hand in Caithness, Sutherland, and Easter Ross. Jarl Sigurd died in 875, and was ultimately succeeded by the scarcely less strenuous Torf-Einar, a son of Jarl Rognvald, and a half-brother of Hrolf (Rollo), the conqueror of Normandy. Einar got his *sobriquet* of " Torf " from the fact of his having learned in Scotland, and taught the islanders, the practice of cutting turf for fuel. He was succeeded by three sons, of whom the two elder, Arnkell and Erlend, fell in the battle of Stanesmoor in England, in 950. The third, Thorfinn Hausacliuf (Skull-Splitter), proved as good as his name, and well maintained the doughty reputation of a family which later, in a collateral line, produced William the Conqueror.

Let us pause here, however, to outline the polity and state of society which had now become established in the Islands. The Orkney Jarls were not autocratic rulers. The Odallers, assembled at the *Thing*, made the laws, on the advice or with the concurrence of the Jarls, but these laws were superimposed on a body of old Norse oral laws which the settlers had brought with them from over-sea. The land law taking no account of primogeniture, the sons of an Odaller succeeded equally to his estate, and a daughter could claim half the share of a son. The eldest son, however, could claim possession of the Bu (English *by*, as in Whitby), or chief dwelling. An Odaller could not divest himself of his odal heritage, except for debt, or in security for a debt, and in such a case a right of redemption lay for all time, not only with his nearest heir, direct or collateral, but on refusal of nearer heirs to avail

Kirkwall

themselves of it, with any descendant whatsoever. Under such a system free men without landed interest actual or prospective were few, and the odal-born formed the bulk of the population. Some of the wealthier Odallers, however, possessed a limited number of thralls, and thraldom was hereditary. Land tax, or *scat*, was paid by the Odallers to the Jarl, and by the Jarl to the King, but in both cases the payment was a fiscal imposition rather than a feudal exaction, the Crown of Norway recognising the obligation of defending the Islands against outside foes in final resort. Apart from this, the overlordship of the mother-country was so slight that in the European diplomacy of the times the Jarls were treated as sovereign princes.

Thorfinn Hausacliuf died *c.* 963 ; and the rule of his five sons, Havard, Hlodver, Ljot, Skuli, and Arnfinn is noticeable for the first of those family feuds which form so marked a feature of the history of the Jarls. Skuli took the title of Earl of Caithness from the King of Scots, and fared against Ljot with a host provided by the King and the Scots Earl Macbeth. Ljot defeated him in the Dales of Caithness, Skuli being slain. Earl Macbeth with a second host, was in turn defeated by Ljot at Skidmoor (Skitton) in Caithness, and here Ljot fell. His other brothers having already disappeared in domestic strife, Hlodver was now left sole. He married Edna, an Irish princess, and their only son Sigurd Hlodverson, the Stout, is one of the most famous characters of the *Saga*. Succeeding his father in 980, Sigurd held Caithness by main force against the Scots. A Scots maormor, Finnleik, the father of the celebrated Macbeth, having challenged him to a pitched battle at Skidmoor by a fixed day, Sigurd took counsel of his mother, for she, as the *Saga* says, " knew many things," that is, by witchcraft. Edna made her son a banner " woven with mighty spells," which would bring victory to those before whom it was borne, but death to the bearer. Armed with this uncanny device, Sigurd defeated his challenger at Skidmoor, with the loss of three standard-bearers. An incident of wider consequence, however, befell Sigurd in the year 995. Olaf Tryggvi's son, King of Norway, came on the Jarl aboard ship in a small bay in the South Isles. The King had the superior force, and, a recent convert himself, he there and then forced christianity upon the reluctant Jarl, and laid him under an obligation to impose the faith upon the people of Orkney. The *Saga* adds, " then all the Orkneys became christian," and so indeed the Islands henceforth remained. Sigurd himself, however, reverted to the old gods, and in the year 1014, he joined the great Norse expedition against Brian, King of Ireland, which led to the battle of Clontarf. In that famous fight the Skidmoor banner again did service, but the spell was broken. " Hrafn the Red," called out the Jarl after two standard-bearers had fallen, and an Icelander who knew its fatal secret had declined to touch it, " bear thou the banner." " Bear thine own devil thyself," rejoined Hrafn. Then the Earl said, " ' Tis fittest that the beggar should bear the bag," at the same time taking the banner from the staff and placing it under his cloak. A little after, the Earl was pierced through with a spear.

Most famous of all the Jarls in the eyes of the Norse was Thorfinn the Great, Sigurd's son by a second marriage with a daughter of Malcolm II, King of Scots. Sigurd, however, was at first succeeded in Orkney by Brusi, Somerled, and Einar, sons of an earlier marriage, while Thorfinn, a boy of five, who had been fostered by the Scots King, was invested by his grandfather in the Earldom of Caithness and Sutherland. On attaining manhood, however, Thorfinn made good his claim to a share of Orkney also, and after many vicissitudes of fighting and friendship with his half-brothers, and with Jarl Rognvald I, Brusi's son, was finally left sole ruler there. He extended his power far and wide over northern Scotland, controlled the Hebrides, and ruled certain parts of Ireland. He even invaded England in the absence of King Hardicanute, and, according to the *Saga*, defeated in two pitched battles the forces that opposed him. In later life Thorfinn visited Rome, and he built a minster, known as Christchurch, in Birsay, the first seat of the Bishopric of Orkney. He died in 1064. Thorfinn's sons Paul and Erlend succeeded, and two years later shared the defeat of their suzerain King Harald Sigurdson (Hardradi) at Stamford Bridge. Returning home by grace of Harold Godwinson, the Jarls ruled in peace for some years. As their sons grew up, however, Hakon, Paul's son, quarrelled with his cousins Erling and Magnus (the future saint), Erlend's sons, and matters grew so unquiet that in 1098 King Magnus Barelegs sent the two Jarls prisoners to Norway, and placed his own son Sigurd over the Jarldom. On the death of King Magnus in 1106, Sigurd returned to Norway to share the vacant throne with his brothers, and the overlords restored Hakon, Paul's son, and Magnus, Erlend's son, to the Jarldom. Quarrels were renewed, with the final result that in the year 1116 Magnus was murdered in the island of Egilsey, where the cousins had

D

met to discuss their differences, by the followers of
Jarl Hakon, Hakon himself more than consenting.
The fame of St Magnus soon spread over the whole
Scandinavian world, and at an early date a church was
dedicated to him even in London. Jarl Hakon, after
the manner of the times, made an expiatory journey to
Rome and the Holy Land, and thereafter ruled Orkney
with great acceptance until his death in 1126. He was
succeeded by his son Paul. While in Orkney in 1098,
however, King Magnus Barelegs had married Gunn-
hilda, a daughter of Jarl Erlend, to a Norwegian
gentleman named Kol. To their son Kali, Sigurd
King of Norway now granted a half share of the
Islands, with the title of Jarl, and from a fancied resemblance to Jarl Rognvald I, insisted on changing
his name to Rognvald. The royal grant being strenu-
ously opposed by Jarl Paul, Rognvald vowed that if
he succeeded in making good his claim, he would erect
a stone minster at Kirkwall, and dedicate it to his
sainted uncle, Jarl Magnus. After many vicissitudes
by sea and land, Rognvald finally proved successful,
and how he fulfilled his vow the Cathedral Church of
St Magnus still shows. St Rognvald—for in 1192 he
too was canonised—was at once the most genial and
the most accomplished of the Jarls, and one of the
great characters of the *Orkney Saga*. He made a famous
voyage to Palestine (1152–1155), fighting, love-making,
and poetising by the way. Incidental to his great
struggle for power with St Rognvald, Jarl Paul had
in 1137 been seized by the famous viking Swein Asleifson

and carried off to Athole, where he was placed in the hands of Maddad, Earl of Athole, who had married his half-sister Margaret. The whole affair is shrouded by mystery, but Countess Margaret appears to have intrigued both with her brother and with Jarl Rognvald

St Magnus' Cathedral, Kirkwall, from South-East

to have her son Harald, a boy of three, conjoined with Rognvald in the Jarldom. In the sequel Jarl Paul mysteriously disappears, murdered, according to one account, at the instigation of the Countess; and in 1139 Jarl Rognvald accepted the young Harald Maddadson as his partner. With occasional intervals of friction this somewhat oddly assorted pair ruled together until 1159, when the checkered career of the genial and many-sided St Rognvald was closed by his assassination in a personal quarrel in Caithness. Thereafter Harald ruled the Islands alone until his death in 1206. A powerful and overbearing man, he quarrelled with John, Bishop of Caithness, blinded the prelate and caused his tongue to be cut out; barbarities which brought King William the Lyon to the borders of Caithness with an army (1202). Harald bought off the King, and on the whole maintained his own in Caithness, although all the circumstances of the times show that a now feudalised Scotland is becoming increasingly able to reassert its authority in these northern parts. Harald got into difficulties with his Norwegian suzerain, King Sverrir, who deprived him of Shetland, which was not again conjoined with Orkney until two centuries later. Harald was succeeded by two sons, David and John, the former of whom died in 1214. Jarl Jòhn, like his father, came into conflict with the Church and with Scotland. Adam, Bishop of Caithness, successor to the mutilated Bishop John, having proved too exacting in the collection of Church dues, the laity appealed to the Jarl, who, however, declined to intervene. Whereupon the outraged laymen burnt the Bishop in a house into which they had thrust him. King Alexander II, came with an army, and not only heavily fined the Jarl, but also had the hands and feet hewn off eighty men who had been present at the Bishop's death. Jarl John was slain in a brawl at Thurso in 1231, and, as he left no son, the line of the Norse Jarls of Orkney ended.

King Alexander II of Scotland granted the Earldom of Caithness, now finally disjoined from Sutherland, to Magnus, second son of Gilbride Earl of Angus, whose wife was a daughter or sister of the late Jarl, and Magnus was also recognised as Earl of Orkney by the King of Norway. The old Jarls had been Norse nobles who held the northern shires of Scotland more or less in defiance of the Scottish Crown, the future Earls of Orkney are great Scottish nobles holding a fief of the Crown of Norway. In consequence, an influx of Scots into the Islands now commenced, which, accelerated by their cession to Scotland in 1468, in the end was the means of transforming the Orcadians into a British community. The history of this Scoto-Norse period is obscure, the Icelandic records largely failing us, but the law of primogeniture being now tacitly applied to the succession, Earl Magnus was followed by two Earls of the name of Gilbride, and the second Gilbride by another Magnus. The latter is mentioned in the *Saga* of King Hakon Hakonson, of Largs fame, as having accompanied the King on that expedition, after which the monarch came back to Orkney with his storm-shattered fleet, and died in the Bishop's palace at Kirkwall on 15th December, 1263. By the treaty of Perth in 1266 Hakon's successor, Magnus the Seventh, ceded to the Scottish Crown all the islands of the Scottish seas, except the Orkneys and Shetland, for an annual payment of 100 merks, to be paid into the hands of the Bishop of Orkney, within the church of St Magnus. Earl Magnus died in 1273, and was successively followed by his sons Magnus and John, the latter of whom, as Earl of Caithness, swore fealty to Edward I of England in 1297. In 1320, however, John's son and successor Magnus subscribed, as Earl of Caithness and Orkney, the letter to the Pope in which the Scottish nobles asserted the independence of Scotland.

Tankerness House and St Magnus' Cathedral

This Magnus was the last of the Angus line of Earls, and was succeeded in 1321 by Malise, Earl of Stratherne, who is supposed to have married his daughter. Malise fell at Halidon Hill in 1333, and his son of the same name succeeded to the three Earldoms of Stratherne, Caithness, and Orkney. Malise II died *c*. 1350, leaving only daughters, and after an unsettled period of

conflicting claims to the succession, Hakon King of
Norway in 1379 finally invested Henry St. Clair of
Roslin, grandson of Earl Malise I, and son-in-law of
Malise II, not only in the Earldom of Orkney, but in
the Lordship of Shetland also. Earl Henry made
himself practically independent of Norway, and built a
castle at Kirkwall in defiance of the Norwegian Crown.
He was succeeded in 1400 by his son Henry, whose active
and interesting career as Lord High Admiral belongs
to the history of Scotland. The Earls of the Sinclair
family lived with considerable state, styling themselves
Princes of Orkney, and their rule was on the whole
popular and fortunate ; but little of outside interest
took place in the Islands at this period. Scottish
customs, however, and traces of Scottish feudal law were
slowly encroaching on the old Norse system. Earl
Henry II's son William was the last Earl of Orkney
under Scandinavian rule. By the Union of Calmar
in 1397 the suzerainty of the Islands had already passed,
with the crown of Norway, to the Kings of Denmark.
In the years 1460–61 a series of raids were—not for the
first time—made on Orkney by sea-rovers from the
Hebrides, and Christian I, King of Denmark, Sweden,
and Norway, failing to obtain redress from the Scottish
Crown for this and other grievances, demanded the
long arrears of the annual tribute payable by Scotland
in respect of the Hebrides, in terms of the treaty of
1266. Charles VIII of France having been called in
as arbitrator, a somewhat ugly contretemps was happily
adjusted by the marriage of Christian's daughter

Margaret to the young King James III of Scotland, and by the terms of the marriage-contract the Danish monarch not only relinquished the quit-rent for the Hebrides, but agreed to pay down a dowry of 60,000 florins. As Christian was able to find only 2000 florins of this money at the time, the Orkneys and Shetland were given in pawn to the Scottish Crown until the balance should be paid, the agreement expressly stipulating for the maintenance of Norse law in the Islands meantime. This happened in 1468, and a few years later King James negotiated the surrender by Earl William St Clair of all his rights in the Islands ; whereupon by Act of the Scottish Parliament the Earldom of Orkney and Lordship of Shetland were in 1471 annexed to the Scottish Crown, "nocht to be gevin away in time to cum to na persain or persains excep alenarily to ane of ye Kingis sonis of lauchful bed," a proviso which went by the board.

The long and painful story of Scottish oppression in Orkney and Shetland has a literature of its own, and can only be briefly referred to here. The Scottish Crown from the first treated the *scats*, in origin and essence a public tax, as a sort of personal perquisite of the King, or part of the Royal patrimonium, and farmed them out, along with the Earldom lands, to one needy favourite or importunate creditor after another, Orkney at the same time being now made liable to all Scottish taxation. Many of these grantees received the jurisdiction of Sheriff, a circumstance which led to the accelerated encroachment of Scottish feudal law on the old Norse legal system. Mere treaty stipulation proved a frail protection to the oppressed Odallers, when the only appeal against strained laws and unjust exactions lay to the Scottish Crown itself, which had installed the oppressors, and whose ministers and judges knew and cared nothing about Odal law. Twice indeed, first in 1503, and again in 1567, the Scottish Parliament expressly recognised the obligation to maintain Norse law, pious or perfunctory opinions which had no practical effect. The two most notorious, because the most powerful, of these Scottish oppressors of Orkney and Shetland were Lord Robert Stewart, whose half-sister, Queen Mary, in 1564 granted him the Sheriffship of both groups, together with all the Crown rights and possessions therein, and Lord Robert's son Patrick. In 1581 Lord Robert was further created Earl of Orkney by his nephew King James VI, and Patrick succeeded him in 1591. Rents and scats being payable to a large extent in kind, by tampering with the old Norse weights and measures, these two harpies in a few years actually increased their revenues from the Earldom one-half. Owing to the unceasing complaints of all classes of the community Earl Patrick was finally imprisoned, and in 1615 executed for high treason. As it had now become apparent that the holders of the Earldom rights had all along simply utilised the local courts and forms of legal procedure for their private advantage, by an Act of the "Lordis of Secret Council," of date 22nd March, 1611, all foreign (*i.e.* Norse) laws theretofore in use in Orkney and Shetland were discharged, and all magistrates in those islands were enjoined to use only " the proper laws of this kingdom." Although this Act was of doubtful validity on more grounds than one, yet it has held good ; and apart from the maintenance of Norse law in its integrity, an ideal which the conditions of the times rendered unattainable, the change was probably the best thing that could have happened for the Islands.

Although the fact had only a transient bearing on the history of the Islands, we must not fail to record that Queen Mary, on her marriage with Bothwell, in 1567, created him Duke of Orkney. After the Queen's defeat at Carberry, Bothwell fled to the Islands, to be repulsed from Kirkwall Castle by the governor. Thence he proceeded northwards to play the pirate in Shetland, and finally to find imprisonment and death in Scandinavia. In the year 1633 the Earldom lands and rights were granted by the Crown to the Earls of Morton, one of whom sold them in 1766 to Sir Lawrence Dundas, with whose descendants, now represented by the Marquis of Zetland, they still remain.

Montrose on his way to invade Scotland in 1650, first landed in Orkney, and the major part of the force with which he met his final defeat at Carbisdale was recruited in the Islands. Under the Commonwealth Cromwell maintained a garrison at Kirkwall, and the Islanders are said to have picked up improved methods of gardening and other domestic amenities from the Ironsides. In the following century the vexed question of Stewart *versus* Guelph brought Orkney its own share of commotion, but the details are, from an historical point of view, of purely local interest.

Of vastly wider import has been the latest appearance of the islands in the arena of history. In that great contest, the sound and fury of which has barely subsided as we write, the Royal Navy found in Scapa Flow an ideal base for the conduct of its widely-spreading operations. And if, as many skilful observers appear to hold, successful strategy at sea proved the slowly-working but inevitably certain cause of the final defeat of the foe, it is not too much to say that for four eventful years the little Archipelago, whose "rough island story" we have here told in outline, stood forth as the pivot of the world's history and of its fate.

The dramatic death of the great Earl Kitchener in Orcadian waters, in June 1916, further riveted the attention of the modern world on a region which to its own sons has never ceased to be the haunted home of Harald the Fair-haired, Olaf Tryggvi's son, and other sea-kings of old.

14. Antiquities

The Orkney Islands offer an especially fertile field to the archaeologist. The sites of a least 70 brochs have been located in the group, as against 75 in Shetland, 79 in Caithness, 60 in Sutherland, and some 70 in the northern Hebrides. These districts constitute the main area of the brochs, which rapidly decrease in number as one proceeds southwards, Forfarshire offering but two specimens, and Perthshire, Stirlingshire, and Berwickshire one each.

"The typical form of the broch," says Dr Anderson,

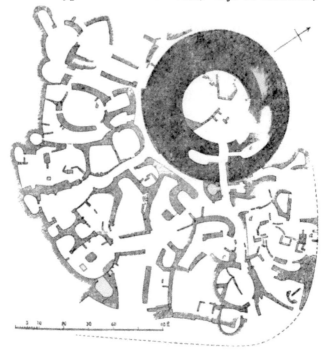

Ground Plan of Broch of Lingrow, near Kirkwall

" is that of a hollow circular tower of dry-built masonry, about 60 feet in diameter and about 50 feet high. Its wall, which is about 15 feet thick, is carried up solid for about 8 feet, except where two or three oblong chambers, with rudely-vaulted roofs, are constructed in its thickness. Above the height of about 8 feet, the wall is carried up with a hollow space of about 3 feet wide between its exterior and interior shell. This hollow space at about the height of a man, is crossed horizontally by a roof of slabs, the upper surfaces of which form the floor of the space above. This is repeated at every 5 or 6 feet of its farther height. These spaces thus form horizontal galleries, separated from each other vertically by the slabs of their floors and roofs. The galleries run completely round the tower, except that they are crossed by the stair, so that each gallery opens in front of the steps, and its farther end is closed by the back of the staircase on the same level.

" The only opening in the outside of the tower is the main entrance, a narrow, tunnel-like passage 15 feet long, 5 to 6 feet in height, and rarely more than 3 feet in width, leading straight through the wall on the ground level, and often flanked on either side by guard-chambers opening into it. This gives access to the central area or courtyard of the tower, round the inner circumference of which, in different positions, are placed the entrances to the chambers on the ground-floor, and to the staircase leading to the galleries above. In its external aspect, the tower is a truncated cone of solid masonry, un-pierced by any opening save the narrow doorway ; while the central court presents the aspect of a circular well 30 feet in diameter bounded by a perpendicular wall 50 feet high, and presenting at intervals on the ground floor several low and narrow doorways, giving access to the chambers and stair, and above these ranges of small window - like openings rising perpendicularly over each other to admit light and air to the galleries."

In some respects the most interesting broch site in Orkney is that at Lingrow, about two miles S.S.E. of

Stone Circle of Stenness

Kirkwall. Contiguous to this building lies a perfect labyrinth of building of secondary occupation. A coin of Vespasian and two of Antoninus were found at Lingrow when the site was excavated in 1870. Another interesting specimen is that at Hoxa in South Ronaldshay. The little that remains of this building is situated on the west side of a small bight which opens northwards on Scapa Flow, and on the east side of the bight stands

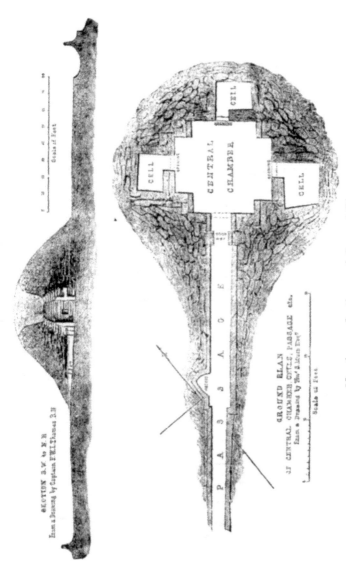

Maeshowe, Section and Ground Plan

the site of a second broch. The two brochs together command both the bight itself and a narrow isthmus of land which connects it with Widewall Bay to the southward. Defence of the land passage between the two areas of water was obviously a prime motive of construction in this case. Other interesting specimens of brochs are at Okstro in Birsay, at Burgar in Evie, and at Burrian in North Ronaldshay ; but many of the Orkney sites remain as yet unexcavated, or only tentatively explored. Stone hand-mills (querns), stone whorls and bone combs, both used in woollen weaving, stone lamps, imitated from Roman models, hair combs, and fragments of pottery are some of the articles found in the Orkney brochs, none of which now show more than a few feet of their original height.

A consideration of the geographical position of the main area of their distribution leaves one with little doubt that they were erected as the best and most economical means that a scanty and scattered population could devise to check continuous piratical incursions, probably directed from Scandinavia and the coastal regions of north-western Europe ; but the exact era of their erection is a question that still awaits definite solution. One may safely conjecture, however, that they were gradually destroyed by the incursions which they were built to withstand.

Other primitive dwellings, of somewhat wider distribution in Orkney than the brochs, are the " Picts' houses," either in the form of chambered mounds, such as the widely-known Maeshowe, or of underground chambers of the type commonly known as weems. Maeshowe (the Maidens' Mound) is situated in the parish of Stenness, near the high road from Kirkwall to Stromness, and its external appearance is that of a truncated cone 92 feet in diameter, 36 feet high, and about 100 yards in circumference at the base. The central chamber is reached by a straight passage 54 feet long, from 2 feet 4 inches to 3 feet 4 inches wide, and from 2 feet 4 inches to 4 feet 8 inches high. The chamber is 15 feet square on the floor, and 13 feet high, the roof being formed by the stones, at the height of 6 feet from the floor, gradually overlapping. The four angles of the chamber are supported by heavy stone buttresses from 8 to 10 feet high, and about 3 feet square at the base. Off the main chamber on the three sides other than that which contains the entrance, are subsidiary cells, the entrances to which are at a height of 3 feet from the floor. The walls of the central chamber are scored with Norse runes some of which tell that the place, which was known to the Norsemen as *Orkahaug*, or the Muckle Mound, was broken into by the Jorsalafarers, or pilgrims who accompanied St Rognvald to Jerusalem (Norse, *Jorsalaheim*) in 1152. Other stones show pictorial designs, several being of fine workmanship. Nothing is known of the date or origin of Maeshowe, which from its elaborate design and peculiarities of detail constitutes a thing unique among the antiquities of Britain. Among weems, or eirde houses, as they are sometimes called, which are irregularly-shaped chambers excavated below the ordinary level of the

E

ground, perhaps the finest Orcadian example is that at Saveroch, near Kirkwall.

At a distance of about a mile and a half west and south-west of Maeshowe is the remarkable series of megalithic monuments collectively known as the Standing Stones of Stenness. The " Ring of Brogar " is

the largest and most complete ; thirteen of its stones are still standing, while the stumps of thirteen more are still *in situ*, and ten are lying prostrate. The highest stone stands about 14 feet above ground, and the total number of stones is calculated to have originally been sixty. This circle has a diameter of 366 feet, and encloses 2½ acres. Of the " Ring

Incised Dragon, from Maeshowe

of Stenness " proper, which has a diameter of 104 feet, only two stones remain erect, but one of these is 17 feet 4 inches high, while another lying prostrate measures 19 feet × 5 feet × 1 ft. 8 inches, and has been calculated to weigh 10.71 tons.

In a valley in Hoy, between the Ward Hill and the precipitous ridge known as the Dwarfie Hammers, lies the celebrated Dwarfie Stone, most puzzling perhaps of all the antiquities of the Islands. This is a huge block of sandstone measuring about 28 feet in length, and varying in breadth from 11 to 14½ feet. Its height falls from 6½ feet at the southern to 2 feet at the northern end. On the west side of the stone is an entrance giving access to two couches or berths, the southern and larger of which has a stone pillow. When first mentioned in writing by Jo Ben in his *Descriptio Insularum Orchadiarum*, in 1529, the Dwarfie Stone was as great a puzzle as it is to-day, and no suggested explanation of its origin or *raison d'être* appears to carry much probability.

Norse Brooch, found in Sandwick, Orkney

Important discoveries of antique jewelry, coins, and other articles of the precious metals have been made in Orkney from time to time. Perhaps the greatest find was that made

Silver Ornaments, found in Sandwick, Orkney

at the Bay of Skaill, in the parish of Sandwick, in March 1858, which consisted of 16 lbs. weight of silver articles, including heavy mantle brooches, torques, bars of silver, and coins, some of the last being Cufic, of the Samanian and Abbasside Kalifs, dating from 887 to 945 A.D., and others Anglo-Saxon.

15. Architecture—(a) Ecclesiastical

The cathedral church of St Magnus at Kirkwall, founded by St Rognvald in 1137 in memory of his uncle, is, except that of Glasgow, the only mediaeval cathedral in Scotland that remains complete in all its parts, and, in the words of the Danish historian Worsae, it constitutes incontestably the finest monument of their dominion that the Scandinavians have left in Scotland.

The church is a cruciform building, comprising nave and nave aisles, choir and choir aisles, north and south transepts, with a chapel off each transept to the east, and a central tower. The interior length of nave and choir is 217 feet 10 inches, exterior length 234½ feet; interior width of choir and aisles at the east end 47 feet 5½ inches; interior width across the transepts 89½ feet; height to vaulting 71 feet; height from floor under central tower to top of weather-cock 133 feet 4 inches. The stones used include two shades of red sandstone, one lilac, one yellow, and one veiny red and yellow, all brought from various parts of Orkney. Sir Henry Dryden professes to trace five distinct styles of architecture in the building, which he dates: 1137–1160, 1160–1200, 1200–1250, 1250–1350, and 1450–1500. All that is beyond conjecture as to the eras of erection, however, is that after St Rognvald's death, in 1159, the building proceeded under the care of Bishop William the Old, until his death in 1168. Part at least of the church was consecrated before 1155, and received the

St Magnus' Cathedral
View from N. Transept looking towards Choir

remains of St Magnus, transferred from Christchurch in Birsay. The general impress of the building is severely Norman.

A conspicuous feature of the church is its great *apparent* size, an effect deliberately produced by a skilful adjustment of proportions, especially by the great relative height, the restricted width, and the spacing of the piers both in nave and choir. Specially fine features of the building are the beautiful east window, and the exquisitely carved doorways in the south transept and west front. These doorways, says Sir Henry, " are probably the finest examples in Great Britain of the use of two stones of different colours in patterns." Unfortunately much of this beautiful parti-coloured work is badly weathered.

The church is at present undergoing " restoration " at the hands of the town council of Kirkwall, who own the fabric, in fulfilment of a bequest by the late George Hunter Thoms, Sheriff of Orkney.

On the highest point of the little island of Egilshay stands a very interesting church, which has been the subject of much archaeological conjecture since its general style appears to combine early Celtic and Norse elements, while it possesses one of those Irish Round Towers the only other Scottish examples of which are at Brechin and Abernethy. Apart from the tower, the church consists of nave and chancel, the nave measuring 30 feet by $15\frac{1}{2}$ feet, and the chancel, which is vaulted, 15 feet by $9\frac{1}{2}$ feet. The tower is now 48 feet high and unroofed, but is said to have been lessened by 15 feet many years ago. An old print shows both church and tower with slate roofs, that of the tower being the conical cap typical of the Irish Round Towers. The church

St Magnus' Church, Egilshay

was used for service down to the early part of last century.

Dryden's conjecture that the church of Egilshay was built by Norsemen after the Irish model shortly after the reconversion of the Islands to christianity *c.* 998

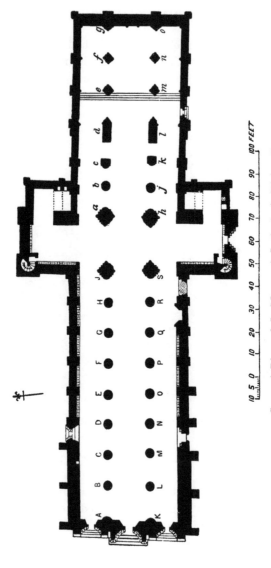

Ground Plan of St Magnus' Cathedral, Kirkwall

A.D., is a view which presents no anomalies either architectural or historical, when we consider that at that date many early Celtic churches must have survived in Orkney, and that, in any case, the close connection between Norse Orkney and the Scandinavian colonies in Ireland would give Orcadians ready access to Irish models.   The alleged dedication of this church to St

**Apse of ancient Round Church,
in Orphir**

Magnus rests on the late and doubtful authority of Jo Ben (1529), but it was no doubt popularly associated with the Saint's name because, as the *Orkneyinga Saga* tells, he entered it to pay his devotions on the day of his murder.

In the parish of Orphir, contiguous to the present parish church, stands the only example in Scotland of a circular church, built in imitation of the church of the Holy Sepulchre at Jerusalem. All that remains of this interesting building, however, is the semicircular chancel, which is 7 feet 2 inches wide and 7 feet 9 inches deep, and about 9 feet of the circular nave on either side. From the curvature of the remaining portion, the diameter of the nave must have been from 18 to 19 feet. Dryden conjectures that the side walls of the nave may have been 15 feet or more in height, and that they were surmounted by a conical roof. This church was almost to a certainty erected by Jarl Hakon, the murderer of St Magnus, after his return from his expiatory pilgrimage to the Holy Land *c.* 1120.

16. Architecture—(*b*) Castellated

The early Norsemen in Orkney built their houses of wood imported from Norway, with only a slight foundation of stone ; and the Norse theory of defence lay in attacking the foe, preferably on sea. Save, therefore, for a few buildings erected by Jarl or Bishop, and by one or two powerful Viking chiefs, stone masonry during the purely Norse period (870–1231) was practically confined to churches. Even during the Scoto-Norse (1231–1468) and Scottish periods men of sufficient power and wealth to erect imposing buildings of stone were still few. Compared, therefore, with an ordinary Scottish county the castellated architecture of the Islands is meagre in quantity ; but several of the buildings of this class that remain are of exceptional interest and beauty.

At the north end of the island of Westray stands

Noltland Castle, the only ancient military structure which Orkney can now show in a state of comparative preservation. The castle is clearly of two periods, but the best authorities now reject the once prevalent idea that the earlier portion could have been erected by

Noltland Castle, Westray (15th century)

Thomas de Tulloch, Bishop of Orkney from 1418 to 1460. The building is supposed to have been added to and beautified by a later Bishop, Edward Stewart, yet so late as 1529 Jo Ben described it as " Arx excellentissima sed nondum completa." Adam Bothwell, the first protestant Bishop, feued the castle to Sir Gilbert Balfour, Master of the Household to Queen Mary, and a doubtful tradition says that Balfour had

orders to prepare the place as a retreat for the Queen
and Bothwell. Finally the place gave refuge to some
officers flying from Montrose's defeat at Carbisdale, and

The Staircase, Noltland Castle

was on that account fired and left dismantled by Cromwell's soldiers. The great staircase of Noltland is one
of the finest in Scotland, excelled only by those of Fyvie
and Glamis Castles. Another special feature of the

castle is the exceptional number of shot-holes, splayed both outwards and inwards, which give its exterior walls the appearance of the hull of an old style man-of-war.

The Earl's Palace at Kirkwall, built by Patrick Stewart, Earl of Orkney, about 1600, is one of the most beautiful specimens of domestic architecture in Scotland, and more regal in appearance than anything north of Stirling and Linlithgow.   Fronting west, the building forms three sides of a square, and is practically entire save for the roof.   On the ground floor is the grand hall, a magnificent apartment about 55 feet long by 20 feet wide, lighted by three splendid oriels, and a triple window in the gable, divided and subdivided by mullions and transoms.   There are two fireplaces in the room, one being 18 feet wide.   At the landing of the main stair there is a beautifully arched apartment of about 9 feet long by 7½ feet wide, commonly styled the chapel, but which may have been a waiting room.   " It is a superb specimen of Scottish seventeenth-century architecture, its oriel windows and turrets being unsurpassed by anything on the mainland, and it is so rich in its details, and spacious in its accommodation, that it is with more than usual regret that one looks on its roofless and decaying walls."   The palace has not been inhabited since towards the close of the seventeenth century.

The Bishop's Palace at Kirkwall, which stands in close proximity to the Earl's Palace and the cathedral, is supposed to have been erected by Bishop Reid (1540–1558), on the site of an earlier building, in which King Hakon Hakonson died in 1263.   A prominent feature of this building is its great length, originally about 196 feet, from which it is supposed that it formed, or was intended to form, one side of a quadrangle. Above what was originally the main entrance is the corbelling of a fine ruined oriel. There is a five-storeyed round

The Earl's Palace at Kirkwall
(c. 1600 A.D.)

tower at the north-west corner of the building, which contains in a niche of its exterior wall a statuette long erroneously believed to represent Bishop Reid. The chief apartment of the palace appears to have been about 26½ feet long by 24 feet wide. Part of the structure is still occupied.

Overlooking the sea, in the extreme north-west of the Mainland, stand the ruins of the Palace of Birsay, erected by Robert Stewart, Earl of Orkney, about 1580, on the site of an old palace of the Norse Jarls. Less elaborate in design, and in a worse state of preservation than the building erected by his son, Earl Robert's palace is sufficiently commodious and noble of outline to make one marvel at the passionate taste for fine buildings of a race which required two such imposing structures in two successive generations. Earl Patrick is said to have partly reconstructed his father's building, modelling his alterations on Holyrood, but the building as it stands more resembles the later courtyard at Dunottar Castle.

17. Architecture—(c) **Municipal and Domestic**

Kirkwall Municipal Buildings, a handsome three-storeyed structure in the Scottish style, erected in 1884, forms the one specimen of municipal architecture in the county that calls for mention. The buildings include a Council Chamber, a Town Hall, a meeting-room for the Commissioners of Supply, a post office, and four or five suites of offices for the burgh officials. The main entrance has a fine semi-classic door-piece, surmounted by two statues of the ancient halberdiers of the burgh in full uniform.

Down to past the middle of the nineteenth century many Orcadian landed proprietors possessed town houses

Town Hall, Kirkwall

at Kirkwall, in which they lived during the winter.
Several of these houses were buildings of some architec-
tural quality, and an interesting survival is Tankerness
House, in the Main Street, the property of Mr Alfred
Baikie of Tankerness.   This building is, with some later
additions or conversions, a quaint conjunction of several

Balfour Castle, Shapinsay

old residences which at one time constituted manses
of cathedral dignitaries.   The house has, however, been
in the possession of the Baikie family since 1641.

Balfour Castle in Shapinsay, the residence of Colonel
W. E. L. Balfour of Balfour and Trenabie, is one of the
finest specimens of Scottish Baronial architecture in
the north of Scotland.   Erected in 1847 from designs
of the late David Bryce, R.S.A., Balfour Castle is

F

effective as a whole without being conspicuous in detail. The house is surrounded by fine gardens and plantations.

Melsetter House in Walls, as it existed down to 1898, is an excellent example of the older Orcadian

Tankerness House, Kirkwall

country houses, few of which now survive. The older part of the building, as it appears on page 32, dates from the later seventeenth century. The house was originally a country residence of the Bishops of Orkney, a fact which may account for its roughly cruciform construction. The place was twice sacked by Jacobites in 1746, during the absence of the owner, Captain Benjamin Moodie, with the Royal army. The gardens and grounds are exceptionally fine for those northern latitudes.

Two other interesting old Scoto-Orcadian mansions are Skaill House in Sandwick, and Carrick House in Eday, the latter erected in the seventeenth century by John, brother of Earl Patrick Stewart, who was himself created Earl of Carrick by King Charles I. At a later date the house was the property of James Fea of Clestrain, who in 1725 captured in the neighbourhood the celebrated pirate John Gow.

In the towns and villages of Orkney and Shetland, where the main street usually runs parallel to the shore, many of the houses stand with one gable towards the street and the other closely overlooking the sea, a feature which gives a distinctly foreign aspect to Lerwick and Stromness in particular. It is sober fact that in many houses in Stromness, granted a taste for fish, and a high tide at the appropriate hour, one can catch one's breakfast from the gable windows before getting out of bed.

18. Communications, Past and Present

Before the days of steam communication there were two ferries across the Pentland Firth to Huna in Caithness, one from Walls and the other from South Ronaldshay. Edinburgh was the usual objective of Orcadians using this route, and Shanks' nag the common means of locomotion, save for persons of quality, who rode horses. The only other means of reaching the Scottish capital, the sole place out of the Islands which old-world Orcadians considered of much account, and a place where all the well-to-do among them had relations, cronies, and " gude-gangin' pleas," was by occasional sailing packets from Kirkwall to Leith.

Steamboat communication between Kirkwall and Leith and Aberdeen was established about 1832, and nowadays there are two boats a week between these ports and the Orcadian capital, besides two to Stromness, and a fortnightly boat to St Margaret's Hope. There is also a daily mail-steamer from Stromness, touching at Kirkwall (Scapa) and South Ronaldshay, to Scrabster in Caithness, thus connecting the Islands with the Highland Railway at Thurso. Stromness is farther connected with Liverpool and Manchester by a weekly steamer, and both Kirkwall and Stromness with Lerwick and Scalloway in Shetland. Kirkwall has communication with the more important of the North Isles, and Stromness with the chief of the South Isles (except South Ronaldshay and Burray) by local steamers sailing several times a week. Motor or horse conveyances run between Kirkwall and Stromness several times a day. There are no railways in Orkney.

The first Orkney Road Act was passed in 1857, and under that and subsequent local Acts, and finally under the Roads and Bridges (Scotland) Act of 1874, the greater part of the roads in the county were constructed. The agricultural depression and land legislation of the " eighties," by attenuating the incomes of proprietors, tended somewhat to check this develop-

ment ; but a dozen years or so later the Congested
Districts Board came to the rescue with grants-in-aid,
largely by means of which many miles of new roads
were completed. With very few exceptions, all the
islands of any population are now amply provided with
good roads.

19. Administration and Divisions

Orkney forms one Sheriffdom with Caithness and
Shetland, and has a resident Sheriff-Substitute at
Kirkwall. There is one Lord-Lieutenant for Orkney
and Shetland, but his deputies and the Commission of
the Peace are appointed separately for each group. In
all other matters of county administration Orkney
forms a separate unit, and for the purposes of the Local
Government (Scotland) Act of 1889 the county is
divided into four districts, one comprising the Mainland,
one the North Isles, one the civil parish of South Ronaldshay and Burray, and the fourth the two parishes of
Walls and Flotta and Hoy and Graemsay. A feature
of county administration perhaps peculiar to Orkney
is that each island forms a unit of assessment for the
construction and maintenance of its own roads. The
control of piers and harbours in Orkney is of a somewhat
diversified order. The Orkney Piers and Harbours
Commissioners, a body acting under special statutes,
control the piers and harbours at Kirkwall, Scapa,
Holm in the Mainland, and at Stronsay, Sanday, and
Westray in the North Isles. Stromness and St Margaret's

Hope possess their own Harbour or Pier Commissioners, and the piers at Longhope, Burray, Egilsay, and North Ronaldshay are managed by the County Council.

Of 21 civil parishes in Orkney, Walls and Flotta, Hoy and Graemsay, and South Ronaldshay and Burray are in the South Isles ; Stromness, Sandwick, Harray and Birsay, Evie and Rendal, Firth, Stenness, Orphir, Kirkwall and St Ola, Holm, and St Andrews and Deerness are in the Mainland ; and Shapinsay, Stronsay, Eday, Rousay, Westray, Papa Westray, Lady, and Cross and Burness are in the North Isles.

Ecclesiastically Orkney forms a Synod of the Church of Scotland, comprising the three Presbyteries of Kirkwall, Cairston (the old name of Stromness), and North Isles. There are 27 ecclesiastical parishes. St Magnus Cathedral is a collegiate church, with ministers of first and second charge.

The standard of education in the Islands is quite up to the average of Scotland. There are higher grade schools at Kirkwall and Stromness, and the ordinary sources of educational revenue are supplemented in various parts of the county by endowments of considerable value.

Orkney joins with Shetland in returning a representative to Parliament.

20. The Roll of Honour

Of heroes of the sword and men of high ruling capacity Scandinavian Orkney produced many, who on a wider

field of action would assuredly have left to the world
at large names now chiefly known to the special student
of Orcadian history. Apart too from mere men of action,
personalities like St Magnus, St Rognvald, and William
the Old, first Bishop of Orkney, would grace the annals
even of a great people. Of distinguished men of the
soldier and statesman type modern Orkney has, however, produced few, a circumstance no doubt in part
due to the distance of the Islands from the centre of
national activity. The Royalist general James King,
Lord Eythin (b. 1589), was a native of Hoy. Fighting
in the Thirty Years' War, in the service of Gustavus
Adolphus, he attained the rank of major-general, but
was recalled to England in 1640. He received a command in the Civil War under Lord Newcastle, and in
1643 was created a peer of Scotland. He died in
Sweden in 1652. Perhaps the two most distinguished
naval officers whom Orkney can lay claim to are Com-
modore James Moodie of Melsetter, who performed
valuable services in the War of the Spanish Succession,
and Admiral Alexander Graeme of Graemeshall, a
veteran of the Great War.

Among explorers two Orcadians stand high, John
Rae and William Balfour Baikie, both explorers of the
highest scientific type, while Baikie was in addition a
good naturalist and a distinguished philologist. John
Rae the famous Arctic explorer, of the Hall of Clestrain,
near Stromness, entered the service of the Hudson
Bay Company, under whose auspices he in 1846 commanded an expedition which linked up the coastline

between the discoveries of Ross in Boothia and those of Parry at Fury and Hecla Strait. In 1847 Rae joined the first land expedition in search of Franklin, and

John Rae

performed important exploration and scientific work. In 1850 he himself commanded a fresh expedition, which mapped a large part of the coast of Wollaston Land, and examined to about 100° the south and east

coasts of Victoria Land.   For the geographical results
of this expedition, and for his earlier survey work, he
in 1852 received the Founders' gold medal of the Royal

David Vedder

Geographical Society. His final Arctic expedition,
organised by the Hudson's Bay Company in 1853,
put the cope-stone to his fame, for on that expedition
he at last succeeded in obtaining definite information

regarding the fate of Sir John Franklin, and in purchasing relics of that hapless explorer and his party from the Eskimos. He died in London in July 1893, at the age of eighty, and was at his own request buried in the churchyard of St Magnus Cathedral at Kirkwall. John Rae was a man of splendid physique and persevering will, and a consideration of his work as a whole places him among the greatest of Polar explorers.

William Balfour Baikie, M.D., was born at Kirkwall in 1825. After serving for some years as a surgeon in the Navy and at Haslar Hospital, he obtained the post of surgeon and naturalist to the Niger expedition of 1854. On the death of the captain of the exploring ship " Pleiad " at Fernando Po, Baikie succeeded to the command, and this voyage, which penetrated 250 miles higher up the Niger than had yet been reached, he described in his *Narrative of an Exploring Voyage up the Niger and Isadda*. In 1857 Baikie left England on a second expedition. In the course of five years he opened up the navigation of the Niger, made roads, established markets, collected vocabularies of various African dialects, and translated parts of the Bible and Book of Common Prayer into Hausa. Baikie died at Sierra Leone, while on his way home on leave, on 17th December, 1864. The loving-care of his fellow-islanders has erected a fitting and touching memorial to him in that great fane of the Isles which holds, or ought to hold, the memorials of all distinguished Orcadians.

Orkney has been the birthplace of many men of high

scientific attainments. Murdoch MacKenzie (d. 1797) was a distinguished hydrographer, who from 1752 to 1771 performed an enormous amount of professional work as surveyor to the Admiralty. His *Treatise on Marine Surveying* is still esteemed. James Copland, M.D., a native of Deerness, was one of the leading medical men of his day (1791–1870). After picking up a knowledge of tropical diseases in West Africa, and travelling in France and Germany, he settled in London. His once famous *Dictionary of Practical Medicine*, however, with its 3509 double-column, small-type pages, a book by one man on every branch of medical science, soon degenerated into one of the curiosities of literature. Thomas Stewart Traill (1781–1862), son of a parish minister of Kirkwall, was professor of medical jurisprudence at Edinburgh University. Himself a man of almost universal learning, he fitly edited the eighth edition of the Encyclopædia Britannica. Matthew Forster Heddle, a son of Robert Heddle of Melsetter, was professor of chemistry at the University of St Andrews, and a mineralogist of great distinction. His great collection of Scottish minerals, now in the Royal Scottish Museum at Edinburgh, is one of the finest things of the kind that any country possesses. To these names we would add those of the Rev. Charles Clouston, minister of the parish of Sandwick, a distinguished meteorologist and naturalist, and of Sir Thomas Smith Clouston, the late distinguished specialist in mental pathology.

Sir Robert Strange, highly renowned as an engraver,

was a son of David Strang, Burgh Treasurer of Kirkwall. Strange joined the Jacobites in 1745, engraved the Prince's portrait and the plate for his bank-notes.   He

Sir Robert Strange, the Engraver

escaped alive from Culloden and evaded the search for him.   Settling in London, he became the foremost in his art.   One of his works is the engraving of West's Apotheosis of George III's children, Octavius and Alfred.

The most celebrated literary man whom the Islands can lay claim to is Malcolm Laing (1762–1818), the Scottish constitutional historian and protagonist in the Ossianic controversy. A friend of Charles James Fox, and a class-fellow of Lord Brougham, Laing sat in Parliament for the county in the Whig interest from 1807 to 1812. In 1808 he withdrew from the literary circles of Edinburgh to his home in the Islands, and here in 1814 he was visited by Sir Walter Scott. Malcolm Laing's younger brother, Samuel, was the author of *Travels in Norway*, *A Tour in Sweden*, and *Notes of a Traveller*. A greater work, however, was his translation of the *Heimskringla*, the old Icelandic history of the early Kings of Norway, by Snorri Sturlason, a performance which had an important bearing on the hero-worship gospel of Carlyle. High literary ability appears to run in the blood of the Laings. Samuel Laing's son, of the same name, was the author of those popular compendiums of nineteenth century science and thought, *Human Origins*, and *Modern Science and Modern Thought*, which are still in wide circulation.

David Vedder (1790–1854), an almost self-taught miscellaneous writer, of large output and considerable excellence, was born in the parish of Deerness. In 1830 he conducted the *Edinburgh Literary Gazette*, supported by De Quincey and others. The various works of Vedder, which include *Orcadian Sketches*, and *Poems, Legendary, Lyrical, and Descriptive*, have fallen on an undeserved oblivion. Vedder is the Orcadian poet *par excellence*. Perhaps the once highly popular novels of the amiable Mary Balfour, or Brunton, *Self-Control*, 1811, and *Discipline*, 1815, would not now

Malcolm Laing
(*From a portrait by Raeburn*)

be considered a passport to fame ; but a later member of the family to which she belonged, David Balfour of Trenabie, has in his *Odal Rights and Feudal Wrongs* written of the Scottish oppressions of Orkney in the sixteenth century in such fine limpid English as makes one regret the purely local interest of his subject. Walter Traill Dennison, who in his *Orcadian Sketch Book* has given his fellow-islanders some light fiction of a high quality, unfortunately elected to write in the strictly local dialect of the North Isles of Orkney, using a phonetic spelling which is a stumbling-block to many readers who are quite at home in ordinary Scots literature. Traill Dennison, like Vedder, is a humourist, and Orkney gifted to the outside world a greater humourist than either when she sent the father of Washington Irving across the sea.

Orkney has produced scarce a churchman or a lawyer who has distinguished himself *as such*. The great Robert Reid, founder of Edinburgh University, was perhaps the most celebrated of her pre-reformation bishops, while James Atkine, a native of Kirkwall, was bishop of Moray in the seventeenth century. William Honyman, of the Honymans of Graemsay, sat on the Scottish bench as Lord Armidale in the early part of the nineteenth century, but his marrying Lord Braxfield's daughter was perhaps his doughtiest feat. The list of Orkney Sheriffs, however, includes the famous names of Charles Neaves, Adam Gifford, and William Edmonstone Aytoun. Aytoun kept house at Berstane, near Kirkwall, and his humourous sketch *The Dreep-daily Burghs*, is supposed to make covert allusion to the climatic and other amenities of the group of parliamentary burghs of which the Orcadian capital at that time formed one.

His connection with the Islands also inspired his poem of *Bothwell* ; and contact with the surrounding seas moved the lively Neaves to verse of nigh as dolorous strain.

21. The Chief Towns and Villages of Orkney

(The figures in brackets after each name give the population in 1911, and those at the end of each section are references to pages in the text.)

Finstown is a small village in the parish of Firth, the half-way house between Kirkwall and Stromness. The village has a pier, an inn, and a monthly cattle fair. (p. 27.)

Kettletoft, a small village on a bay of the same name, on the east side of Sanday, is the business centre of the island, and a minor herring-fishing station.

Kirkwall (3810), is situated at the northern end of a low isthmus dividing the Mainland into two, and has a harbour on two sides, that at Scapa Bay to the south being the port of call for the daily mail-steamer to Caithness. The harbour of Kirkwall proper, facing to the north, was designed by Telford, and is the place of general traffic. The town takes its name, in Norse *Kirkjuvágr*,[1] churchbay, from the old parish church of St Olaf, supposed to have been erected by Jarl Rognvald I, *c.* 1040, but it only became the capital of the Islands after the foundation of the cathedral in 1137, and the consequent transfer of the

---

[1] *Kirkjuvágr* softened into Kirkwa, and Scots mistook the final syllable for their own *wa'* =wall ; hence Kirkwall.

Stromness, Orkney, about the year 1825
(From an old print)

G

seat of the Bishopric from Birsay. Kirkwall received its earliest charter as royal burgh from James III, in 1486, and was visited by James V, in 1540. Of several later charters the governing one is that granted by Charles II, in 1661. The population consists mainly of professional men, tradesmen, artisans, and labourers, there being no manufactures of consequence, although fish-curing is prosecuted to some extent, and there are two distilleries adjacent to the town. The volume of trade is very considerable, however, as the town shares with Stromness the position of shopping and shipping centre for a large and comparatively well-to-do agricultural community. The ancient Grammar School of Kirkwall, now housed in a spacious modern building and styled the Burgh School, has given the rudiments of education to many distinguished sons of town and county. The institution dates at least from Danish times (1397-1468). Owing to the narrowness of the streets, and the many old buildings both public and private, the general impression created by the town is one of quaint antiqueness ; while the Oyce, a miniature inland sea on its north-western outskirts, imparts an additional touch of picturesque individuality to the place. Two weekly newspapers, *The Orcadian* and *The Orkney Herald*, are published at Kirkwall. (pp. 16, 27, 41, 42, 50, 53, 55, 58, 68, 77, 79, 81, 84, 85, 86, 90, 95.)

Longhope, in Walls, although it contains no village, has scattered along the lower coasts of its splendid bay a Customs House, a good hotel, two post offices, and a few shops. Two martello towers, one on each side of the entrance to the bay, attest the importance of the anchorage at the period of the Great War, when as many as 200 sail of merchantmen at times lay there awaiting convoy. The bay is now a coaling station and much appreciated harbour of refuge for steamers, especially trawlers on their way to and from the northern fishing grounds. It also constitutes a sort of headquarters for the naval vessels which now annually visit Scapa Flow for exercise. (pp. 26, 27, 43.)

Pierowall is a small village on the bay of the same name, on the east side of Westray. The bay is a port of shelter for trawlers, and the village the trading centre for the island.

St Margaret's Hope (400) is a picturesque village situated on a bay of the same name at the northern extremity of South Ronaldshay. The bay, named from an early church dedication to St Margaret of Scotland, was known to the Norse as Ronaldsvaag, and was the rendezvous of King Hakon's fleet in the expedition of Largs. The village is after Kirkwall and Stromness the chief trading centre in Orkney. (pp. 27, 84, 85.)

St Marys is a fishing village situated on Holm Sound, 6 miles S.S.E. of Kirkwall. From here Montrose sailed to Scotland on his final venture in 1650.

Stromness (1653), one of the most picturesquely situated towns in Scotland, skirts the W. and N.W. sides of a beautiful bay in the S.W. of the Mainland, which for years was the last port of call for outward bound Greenland whalers and the vessels of the Hudson's Bay Company, and also a touching point both outwards and homewards for the ships of many Arctic expeditions. As a burgh of barony Stromness dates from 1817, but as a village it had already made history. In the early part of the eighteenth century the vessels engaged in the American rice trade having made the safe little bay the depôt for unloading their cargoes for the various ports of Britain, the growing commerce of the place attracted the jealousy of Kirkwall. The magistrates of the Orcadian capital, founding on two obscure Acts of 1690 and 1693, which appeared to confine the right to such trading to freemen inhabiting royal burghs, and to such other communities as agreed to pay cess to a royal burgh, tried to tax the trade of Stromness out of existence. The people of Stromness, after submitting to these imposts for many years, between 1743 and 1758 fought their oppressors in the Court of Session and the House of Lords, both of which tribunals decided that " there was no sufficient right in the burgh of Kirkwall to acssss the village of Stromness, but that the said village should be quit therefrom, and free therefrom in all time coming." As the case was a test one, the courage shown by this little Orcadian community freed all the villages of Scotland from the exactions of the royal burghs. Stromness was visited by Sir Walter Scott in 1814, and by Hugh Miller in 1847, the latter of whom exhumed from the neighbouring flagstones the specimen of asterolepis which he deals with in his *Footprints of the Creator*. Interesting natives or inhabitants of Stromness were John Gow, or Smith, the original of Scott's Cleveland in *The Pirate*, a man whose true history has been told by Defoe ; and George Stewart, the " Bounty " mutineer, of whom Byron writes in his poem *The Island* :

> And who is he ? the blue-eyed northern child
> Of isles more known to men, but scarce less wild,
> The fair-haired offspring of the Orcades,
>   Where roars the Pentland with its whirling seas.

Stromness is a place of considerable trade, and has the advantage over Kirkwall of direct communication with Liverpool and Manchester. There is a good museum in the town. (pp. 11, 27, 41, 42, 83, 84, 85, 86, 87.)

Whitehall (200), a village on the shores of Papa Sound, near the northern extremity of the island of Stronsay. (p. 41.)

SHETLAND

By T. MAINLAND

PREFACE

THE author begs to acknowledge his indebtedness to Messrs Peach and Horne (in Tudor's *Shetland*) for the notes on Geology in Chapter 5; to Mr John Nicolson author of *Arthur Anderson,* for the Notes and photograph in Chapter 18; and to Goudie's *Antiquities* for historical information.

SHETLAND

1. County and Shire.[1] Name and Administration of Shetland

Shetland or *Zetland* is derived from the Norse *Hjaltland,* variant spellings being *Hieltland, Hietland, Hetland.* The Norse word is of doubtful origin.

Like Orkney, Shetland cannot be regarded as strictly a Scottish county till the seventeenth century. During that century various County Acts were framed for the better government of the islands. These lasted till 1747, when heritable jurisdiction was abolished and the judicial administration of Shetland was assimilated to the general Scottish system.

Shetland has the same Lord-Lieutenant as Orkney, but separate Deputy Lieutenants and Justices of the Peace. It shares a Sheriff-Principal with Orkney and Caithness, but has a Sheriff-Substitute of its own. The County Council, the Education Authority, Parish Councils, and Town Council of Lerwick are the administrative bodies. The civil parishes are : Unst, Fetlar, Yell, Northmavine, Delting, Walls, Sandsting, Nesting, Tingwall, Lerwick, Bressay, Dunrossness.

In Parliament one member represents both Orkney and Shetland.

Ecclesiastically the parishes are divided into three

[1] See p. 1.

presbyteries, Lerwick, Burravoe and Olnafirth, which make up the Synod of Shetland.

2. General Characteristics

The county of Shetland is entirely insular, and its characteristics are varied. The coast-line is generally broken and rugged, and in many places precipitous ; while the larger islands are intersected by numerous

Crofter's Cottage

bays and voes stretching far inland, which form safe and commodious places of anchorage and easy means of communication. No point in Shetland is more than three miles from the sea. Detached rocks and stacks, some high above the water and others below the surface, present a forbidding aspect to the spectator, and increase the dangers of navigation round the coast.

To be near the sea—their chief source of food—the early settlers made their homes close to the shore, where also they generally found the soil better than further inland. Accordingly most of the houses and crofts are situated along the coast, and in particular at the heads of voes, where often townships and villages have been formed. One striking feature is the bold contrast between the green cultivated township and the dark background of moor or hill, sharply marked off from each other by the ancient " toon " dykes. Still more noticeable are the large tracts of permanent pasture, where one may still see the ruins of cottages, once the homes of crofters who were evicted to make room for sheep.

The principal fuel in the islands is peat, the cutting of which, carried on for centuries, ever since the time of Torf Einar, who is said to have taught the natives the use of " turf " for burning, has denuded the surface, and left it blacker than it otherwise would be. That, and the practice of " scalping " turf for roofing purposes, have laid bare great stretches of what might have been fairly good pasture ground.

The proportion of arable land is very small compared with the whole land surface of the islands. For this reason Shetland has never figured as an agricultural county ; and other causes may be mentioned, such as divided attention between the two branches of industry —fishing and crofting ; indifferent soil ; variable climate ; antiquated methods of cultivation ; and want of markets for agricultural produce.

Being far removed from the centres of commerce, and having few natural resources in themselves, the islands are almost entirely lacking in manufactures. Fishing and farming are the chief industrial pursuits, while the rearing of native sheep, ponies and cattle form a considerable part of the rural economy. Along with every croft or holding goes the right of common pasturage in the hills and moors, called " scattald." Crofts would be of little value without this common grazing, which is a remnant of the old Norse odal system of land tenure, and is peculiar to the islands of Shetland and Orkney. The word *scattald* is from *scat*, a payment, akin to *scot* in " scot and lot."

3. Size. Position. Boundaries

In size, Shetland ranks fifteenth among the Scottish counties ; but in population only twenty-seventh. The total area (excluding water) is 352,319 acres or 550½ square miles. The group consists of about one hundred islands, of which twenty-nine are inhabited. The largest is Mainland, which embraces about three-fourths of the whole land-surface ; and the others, in point of size, are Yell, Unst, Fetlar, Bressay and Whalsay. Yell measures 17½ miles by 6½, Unst 12 by nearly 6. The smaller islands include East and West Burra, Muckle Roe, Papa Stour, Skerries, Fair Isle and Foula ; the remainder consisting of islets, holms, stacks and skerries. From Fair Isle, which lies about mid-way between Orkney and Shetland, to Sumburgh Head, is 20¼ miles ;

and from that point to Flugga in Unst, in a straight line, is 71 miles. From Sumburgh Head to Fethaland, the most northerly part of the Mainland, is 54 miles; and the greatest breadth, from the Mull of Eswick to Muness, is about 21 miles.

Including Fair Isle and Foula, the group lies between the parallels of 59° 30′ and 60° 52′ north latitude; and between 0° 43′ and 2° 7′ west longitude. The general trend of the islands is north-easterly, and their position in relation to the mainland of Scotland is also in that direction; Sumburgh Head being 95 miles north-east of Caithness and 164 miles north of Aberdeen.

The North Sea washes the eastern sea-board, and the Atlantic Ocean the western; sea and ocean join forces between Sumburgh Head and Fair Isle to form that turbulent tideway called the "Roost." They also meet north of Yell and Unst, where the Atlantic sets strongly through Yell Sound and Blue Mull Sound for six hours during flood tide, and the North Sea, in its turn, in the opposite direction for other six hours.

4. Surface and General Features

The surface of the larger islands is hilly, and the general direction of the ridges is north and south, corresponding to the length of the islands. The inland parts present an undulating surface of peat bogs and moorland, dotted here and there with fresh-water lochs, which accentuate, rather than relieve, the monotony of the landscape. The hills are round or conical in shape and of moderate height, ranging from 500 to 900 feet. Unlike the rounded hills of the Lowlands of Scotland, which are often grassy to the top, most of the Shetland hills are brown and barren, being covered with heather and moss or short coarse grass. Neither have they the rugged grandeur of the Highlands, for the sides are generally smooth and rounded, the result of intense glaciation.

The highest in the Mainland is Ronas Hill, North-mavine, a great mass of red granite, rising to a height of 1475 feet, forming the culminating point of the North Roe tableland, and having the hilly ridge of the Björgs on its eastern side. A long range of hills extends through the parish of Delting, from Firth Ness to Olnafirth ; and from the south side of that inlet it extends south-ward to Russa Ness on the west side of Weisdale Voe. This is the loftiest range, and the southern ridge forms a complete barrier between the east and west side of that part of the Mainland. The highest hill in this chain is Scallafield, 921 feet. Other two parallel ridges lie east of this, the one from the east side of Weisdale Voe to Dales Voe in Delting ; and the other from near Scalloway in a north-east direction to Collafirth. There are breaks in these hilly ridges, notably where the county road crosses over near Sandwater, and again between Olnafirth on the west and Dury Voe on the east. The other range worthy of note extends from the Wart of Scousburgh in Dunrossness, with intervening breaks, to the east of Scalloway, where it divides into two ridges, and stretches onward to the east and west

sides of Dales Voe. Fitful Head in the south, 928 feet ;
Sandness Hill in the west, 817 feet ; the Sneug in Foula,
1372 feet ; the Wart of Bressay, 742 feet ; and the Noup
of Noss, 592 feet, are conspicuous land-marks round
the coast. The island of Yell is hilly, with parallel
ridges running north-east and south-west—highest point,
the Ward of Otterswick. A range of hills extends along
the west side of Unst—highest point, Vallafield, 708 feet.
In the extreme north is Hermaness, 657 feet ; while on
the opposite side of Burrafirth is the still higher hill
of Saxa Vord, 934 feet.

In the larger islands the land is generally lower on
the east side than on the west ; and in many cases the
hills are steeper on the west than on the east. Most
of the smaller islands are low-lying and grassy, forming
excellent pasture ground for sheep.

There are many fresh-water lochs in the islands, the
largest being Girlsta Loch in Tingwall, Loch of Cliff in
Unst, and Spiggie in Dunrossness. They are nearly
all well stocked with trout, which makes Shetland
attractive to anglers. Shetland has no streams large
enough to be called rivers ; but it has burns in plenty,
rippling brooks in summer and brawling torrents of
brown peaty water in winter, when the soil is sodden
with rain or snow.

5. Geology and Soil

The geological map shows that metamorphic rocks
cover the greater part of Shetland. These rocks are

represented by the clayslates and schists that extend from Fitful Head to the Mull of Eswick ; and by the gneiss found from Scalloway to Delting, and also on Burra Isle and Trondra, Whalsay and Skerries, Yell, the west side of Unst and of Fetlar, the east side of North-mavine, and the north sea-board of the Sandness-Aithsting peninsula. Associated with the two series of metamorphic rock are bands of quartzite ; while at Fladdabister, Tingwall, Whiteness, Weisdale, North-mavine and Unst are beds of limestone.

The geological formation of Unst is interesting. The gneiss on the west is succeeded by a band of mica, chlorite and graphite schists. Next come zones of serpentine and gabbro, to be followed by schistose rocks at Muness. Fetlar shows a similar formation.

Of the igneous rocks we may mention first the intrusive granites of Northmavine, Muckle Roe, Vementry, Papa Stour, Melby and South Sandsting. A bed of diorite extends from Ronas Voe to Olnafirth. Part of Muckle Roe also shows diorite. Syenite occurs round Loch Spiggie and is traced northward through Oxna and Hildasay to Bixter and Aith. Lava, tuff, and other volcanic materials appear in various parts of the islands, as between Stennis and Ockran Head, in Northmavine, Papa Stour, the Holm of Melby, Vementry and Bressay.

The predominating sedimentary rocks in Shetland belong to the Old Red Sandstone formation. These are found from Sumburgh Head to Rova Head, and in Fair Isle, Mousa and Bressay. The fault or boundary line between them and the metamorphic rocks is clearly traceable at various points along the east of the Mainland. Foula, Sandness and Papa Stour, on the west, show remnants of Old Red Sandstone. This formation is believed to have once covered the whole area now represented by the Shetlands. To the same geological period the hardened sandstone is considered to belong, which occurs in the western peninsula, covering most of the district of Sandness, Walls, Sandsting and Aithsting.

Glacial phenomena—striae, moraines, boulders—are interesting features in the geology of Shetland. Over Unst, Fetlar, Yell and the north of Mainland the movement of the ice had been uniformly from east to west. Over the middle and southern districts, the glaciers had curved in a north-westerly direction.

The variety of rocks accounts for the variety of soils in the islands. There are many fertile regions of sandy and loamy soil. Perhaps the best is the limestone district of Tingwall. Most of the county, however, is covered with peat moss and peaty soil.

6. Natural History

The mammals of Shetland are comparatively few. There are no deer, foxes or badgers. Rats and mice are numerous, hedgehogs are fairly common. Moles and bats are unknown. So, too, are snakes, lizards, frogs and toads. The weasel is found in many districts, and its near relative, the ferret, is used for hunting rabbits, which are plentiful everywhere. Hares were imported some time ago, but have a hard struggle for existence with so many enemies around. Along the shore may occasionally be seen the sea-otter, when he comes up on a rock to enjoy a feast of sea-trout, or to

Kittiwakes. Noss Isle

venture inland in search of fresh water. He must needs be wary, as he is greatly sought after for the sake of his valuable skin. Another amphibian is the seal, whose habitat is the base of inaccessible cliffs and out-lying rocks and isles.

The birds of Shetland may be divided into residents, summer visitors, and winter visitors. In the great annual migrations which take place in spring and again in autumn, the islands form a resting-place for the feathered voyagers, some of whom stay to nest. Sea fowl naturally predominate. The gull family is the

Shag on Nest.   Noss Isle

most numerous and includes the great black-backed gull, the lesser black-back, the herring gull, the common gull, the black-headed gull and the pretty kittiwake. Shags (scarfs) and cormorants are abundant all the year round ; and in their season come the guillemots, razorbills, puffins, manx shearwaters and little auks. The guillemots congregate in thousands on the rocky

H

ledges of cliffs, where each female lays a solitary egg on the bare rock. The egg is so formed that, even if disturbed, it will not roll off the shelf on which it lies. The plumage of the black guillemot or " tystie " during the winter season is of a mottled-grey colour, and the bird is difficult to recognise in its changed appearance. Eider ducks are abundant, and the great northern diver (emmer goose) may often be seen.

Colonies of fulmar petrels are spreading all round the coast, while the stormy petrel comes to nest in the outer islands. The piratical skua is common ; and the great skua, which is strictly protected, is on the increase in certain localities. Among the shore birds may be mentioned the noisy terns and oyster-catchers. In the winter come the turnstones, sandpipers, dunlins and redshanks. Among the divers may be noted the golden-eye, merganser, and long-tailed duck, while the stock-duck, teal and widgeon are fairly plentiful.

Wild geese and swans, and flocks of rooks occasionally rest on their journey, but the rooks, though numerous elsewhere in Britain, do not take up residence in Shetland longer than they can help. Ravens and hooded-crows are plentiful. The peregrine falcon, the merlin, the kestrel and the lordly sea-eagle are among the birds of prey. Of game birds, the ringed plover, curlew, snipe, golden plover and rock pigeons are common, while woodcock make their appearance during the migratory flight in autumn.

Starlings, sparrows and linnets are plentiful ; and owing to the almost entire absence of other songsters, the song of the skylark is more noticeable. Among summer visitors may be mentioned the wheat-ear, land-rail and peewit, while swallows and martins may be seen occasionally. Winter visitors include field-fares, buntings, redwings, robin redbreasts, blackbirds, thrushes and the smallest of all British birds, the golden-crested wren. The tiny wren, misnamed the " robin," with its cheery song and hide-and-seek ways among the rocks, stays all the year round, as do also the rock and meadow pipits.

Absence of trees and hedgerows, and the general bareness of the ground, bring into prominence the different varieties of wild flowers. On the moors and hillsides grow the crisp heather and heath, with occasional patches of crowberry, while peeping out among these are the milk-wort, the butter-wort and bog asphodel, with the downy cotton-grass in damp places. The pretty yellow tormentil is everywhere among the stunted grass. The spotted orchis, the yellow buttercup and the lovely grass of Parnassus are also conspicuous on the grassy uplands. In meadow lands and marshy places are found the purple orchis, marsh-marigold, marsh cinquefoil, lady's smock, ragged robin, and red and yellow rattle, while in ditches may be seen different varieties of the crow-foot tribe and scorpion grass (forget-me-nots). On dry banks may be found the scented thyme, white and yellow bed-straws, and near by, the bird's foot trefoil. Daisies, scentless Mayweed and eyebrights are everywhere. In some corn-lands the red poppy shines forth ; while the wild mustard and radish (runchie) are only too common. Growing in the cliffs are rose-root, scurvy grass and sea-campions (misnamed sweet-william). The vernal squill, hawk-weed and sheep's scabious spangle the green pasture, and sea-pinks are all round the shore, extending in some places to the water's edge. The mountain-ash, the wild rose and the honeysuckle grow in sheltered nooks.

7. Round the Coast—(a) Along the East from Fair Isle to Unst

The extent of coast-line is enormous, and only the outstanding features can be noted here.

Fair Isle, rock-bound, precipitous and lonely, with but one or two small creeks where vessels may shelter, has two lighthouses, each with a fog-siren and a group-flashing white light—the Scaddon, visible 16 miles, and the Scroo, visible 23 miles. It was on Fair Isle that *El Gran Grifon*, one of the Armada ships, was wrecked in 1588

Sumburgh Head, the most southerly point of the Mainland, is also capped with a lighthouse, perched 300 feet above the swirling eddies of the Roost. Erected in 1820, it was the first lighthouse on the Shetland coast. Its group-flashing white light is visible 24 miles in clear weather. North of Sumburgh Head are Grutness Voe and the Pool of Virkie, with their low-lying sandy shores ; while near by are Sumburgh House and the ruins of

Jarlshoff. From this point northward the coast is rocky, with moderately low cliffs, which are broken up by the inlets of Voe and Troswick. The next conspicuous point is the headland of Noness, crowned by a small, quick-flashing white light, a guide to the busy fishing centre of Sandwick, and the fine sandy bay of

Sumburgh Head and Lighthouse

Levenwick, which lies directly opposite. Passing through Mousa Sound, we have on the right the low-lying grazing island of Mousa with its famous Pictish castle, and on the left Sandlodge, near which are disused copper-mines. Two miles north of Mousa is the low-lying point of Helliness, having at its base the small harbour of Aithsvoe. From here onwards the coast is of varying heights, and broken by the exposed creeks of Fladdabister, Quarff, Gulberwick, Sound and Brei Wick. Bressay lighthouse, which has a revolving red and white light, visible 16 miles, shows the way to Bressay Sound and the harbour of Lerwick.

Bressay Lighthouse and Foghorn

The shores of Bressay are low-lying on the west and north ; but on the south are the high cliffs of the Ord and Bard, the latter having the Orkneyman's Cave and the mural arch of the Giant's Leg at its base. On the east side of Noss Isle the cliffs are also high, forming the favourite breeding-ground of myriads of sea-birds. The diversity of the coast-line, with the sudden transition

Cliff Scenery, Noss. Bressay

from an uninteresting shore to bold, precipitous cliffs, is owing to the dip of the rock-strata. If the dip is to the sea, there is a more or less gradual slope *to* the shore ; but if the dip is *from* the sea, then there is the towering cliff, facing the waves like a giant wall. A good example of this is in the island of Noss, which is 592 feet high at the Noup, and gradually slopes down to near sea-level at the western side. The general outline of the cliffs, too, depends on the texture of the rocks. With sandstone, they tower up in regular layers like a wall, bold and massive, with clear-cut caves and tall stacks. With schistose and granite rocks, however, the cliffs are more broken and rugged.

Proceeding from the north entrance of Lerwick harbour, we find a light on Rova Head, with red, white and green sectors, guiding the mariner to and from this narrow channel. The sea is dotted with a number of rocks and holms, requiring a skilled pilot to negotiate. Towards the west open up the fine bays of Dales Voe, Laxfirth, Wadbister Voe and Catfirth ; and off the point of Hawksness is the Unicorn rock, on which the ship that was chasing the Earl of Bothwell was wrecked (1567).

Facing the bold promontory of the Mull of Eswick stands the Maiden Stack ; to seaward the Hoo Stack and Sneckan ; and extending eastward a succession of rocks and skerries, which go by the name of the " stepping stones." Passing the bay of South Nesting, we approach the " bonnie isle " of Whalsay, with the fine bay of Dury Voe opposite. Between Whalsay and the land lie a number of isles, the largest being West Linga, while from the outer side, islands and rocks stretch in an easterly direction, the farthest group being called Out Skerries. The largest of these, Housay, Bruray

Mavis Grind, looking south

and Grunay, are inhabited ; while on a small isle, called Bound Skerry, is erected a fine lighthouse, which shows a bright revolving white light, visible a distance of 18 miles round. Passing Vidlin Voe and doubling the long promontory of Lunna Ness, we reach Yell Sound. On the right lies the island of Yell ; on the left the triple openings of Swining Voe, Collafirth and Dales Voe ;

and further on Firths Voe and Tofts Voe, with the ferry of Mossbank between. After Orka Voe, we round Calback Ness and enter the fine bay of Sullom Voe, extending about eight miles inland. Only a narrow neck of land separates it from Busta Voe on the west, while a little to the northwards, at Mavis Grind, the distance between Sullom Voe and the Atlantic is only a stone's throw. After Gluss Voe and the pretty bay of Ollaberry, Colla-firth is reached, and then Burravoe, the farthest road-stead in Yell Sound. Looking back from this point, we view the whole Sound with its many isles and holms, low-lying, and affording good pasture for sheep. The largest are Lamba, Brother Isle, Bigga, and Samphrey. Near Samphrey are the Rumble Rocks, with a beacon to aid navigation.

Returning to Yell Sound, we find on its north side the two harbours of Hamnavoe and Burravoe, in Yell. Proceeding north through the Sound, we pass the bay of Ulsta and the Ness of Sound. From Ladie Voe to Gloup Holm the coast is bold and rocky, with only one inlet, Whale Firth, which runs inland till it almost meets Mid Yell Voe. In the extreme north is Gloup Voe. Blue Mull Sound, which separates Yell from Unst, has the harbour of Cullivoe near the middle and Linga Island at the south entrance. On the east coast of Yell are the fine bays of Basta Voe and Mid Yell Voe, with the island of Hascosay between.

Across Colgrave Sound lies Fetlar, one of the most fertile of the islands. It is elevated towards the north-east, where Vord Hill rises 521 feet. Hereabout the coast is precipitous and very imposing. The principal openings are Tresta Wick in the South and Gruting Wick in the east. Brough Lodge is a calling-place for steamers.

Unst is the most northerly part of the British Isles.

Muckle Flugga Lighthouse
(Most northerly part of the British Isles)

In the south is the harbour of Uyea, east of which is Muness with Muness Castle. The principal harbour in Unst is Balta Sound, completely protected by the islands of Balta and Huney. Further north is Haroldswick, named after Harold Fairhair of Norway, who

landed here on his expedition against his rebellious
subjects.   After passing the sandy bay of Norwick and
rounding Holm of Skaw, we reach the northern extremity
of Unst.   The heights of Saxa Vord and Hermaness
hem in the entrance to Burrafirth.   Rising from a
group of rocks, about a mile to seaward, is the Flugga,
with a fine lighthouse, which shows a fixed light, with
white and red sectors.   From Hermaness to Blue Mull
Sound, the west of Unst is bold and rocky, with only
one inlet of importance, Lunda Wick.

8. Round the Coast—(b) Along the West from Fethaland to Fitful Head

A mile or two off the point of Fethaland are the
Ramna Stacks, huge rocks like giant sentinels guarding
the northern extremity of the Mainland.   From this
point to the isle of Uyea, and onwards to Ronas Voe,
the chief feature of the coast is the high and rugged
granite cliffs, which gradually increase in height till
the voe is reached.   This fine natural harbour, lying
round the base of Ronas Hill, forms, with Urafirth
on the opposite side, an extensive peninsula to the
westward.   The coast line of the peninsula is rugged
and much indented, with wonderful caves and sub-
terranean passages, wrought out by the action of the
sea, the Grind of the Navir and the Holes of Scradda
being the most remarkable.   Off the south coast lie

the famous Dore Holm with its natural arch, and the Drongs, resembling a ship under sail. Hillswick, in Urafirth, is the terminal port of call for steamers on the west side, and forms a convenient centre for the many fine trout-fishing lochs in the neighbourhood.

On the east of the extensive bay of St Magnus lies Muckle Roe, with its sea-face of red granite cliffs. To the south are the smaller grazing islands of Vementry and Papa Little, with the channel of Swarbacks Minn between, forming the common entrance to Busta Voe, Olnafirth, Gonfirth and Aith Voe. Between the last opening and Bixter Voe, on the opposite side, runs an isthmus which connects Aithsting, Sandness, Walls and Sandsting with the other part of the Mainland. The northern side of the peninsula thus formed is pierced by the openings of Clousta Voe, Unifirth and West Burrafirth, while lying off the north-west extremity is the fertile island of Papa Stour, with its high cliffs and beautiful caves, Christie's Hole being the finest. About three miles to seaward are the dangerous rocks of Ve Skerries. The coast from Sandness to Vaila is bold and rocky, the headland of Watsness being the turning-point on this rugged coast. About eighteen miles to the south-west lies the lofty island of Foula. The east side of the island is comparatively low-lying, but the land rises towards the west, where the cliffs tower 1220 feet above the sea. A dangerous shoal, the Haevdi Grund, lies to the east of the island, and it was here that the liner *Oceanic* was wrecked (1914).

The island of Vaila lies across the mouth of Vaila

Sound, and bounds the entrance to the extensive bay of Gruting Voe. From Skelda Ness to Scalloway the distance is about seven miles across an arm of the sea, which runs north and is broken up into many openings. Taken in order, these are Skelda Voe, Seli Voe and Sand Voe; Sand Sound, the common entrance to the voes of Bixter

The Kame, Foula

and Tresta; Weisdale Voe; Stromness Voe with the tidal loch of Strom; and Whiteness Voe. Lying off the entrance of these two are the islands of Hildasay, Oxna and Papa. Opposite Scalloway, and forming a shelter to its harbour, is the island of Trondra. Further south are East and West Burra and Havra, with the long channel of Cliff Sound on their eastern side. The fertile island of St Ninian, locally known as St Ringans, is joined to the land by a narrow isthmus of sand. To the south lies the isle of Colsay opposite the inlet of Spiggie. The high cliffs of Fitful Head form the termination of the west, as Sumburgh Head of the east. Between lies the fine sandy bay of Quendale, with its background of sand dunes and rabbit warrens, while to the south are the islands of Lady's Holm, Little Holm and Horse Island.

9. Climate

As Shetland is entirely insular, and surrounded by ocean currents considerably warmer than those of other places in the same northern latitude, the climate is wonderfully equable, extremes of heat and cold being rare. The prevailing winds come from the south and west, laden with warm moisture from the Atlantic to temper the atmosphere. Added to that is the general " drift " of the Atlantic towards the British Isles. Of this warm current, Shetland gets its share.

When the air becomes laden with moisture, it is lighter than dry air, and in consequence the barometer is low. Wet and stormy weather usually follows. When the air is clear and dry, then the barometer is high, and good weather may be expected. The former is called cyclonic, and the latter anti-cyclonic types of weather. The atmospheric pressure varies as the storm moves onward ; hence there is usually a falling barometer till the storm is over, succeeded by a rising barometer. The cyclone, as the storm is called, has two motions, a circular motion opposite to the hands of a clock, and a general forward movement as a whole, usually from some point west to some point east. In winter the deficiency of atmospheric pressure in the neighbourhood of Iceland accounts for the procession of westerly gales from the Atlantic during that season, causing the southerly type of wind to prevail over Shetland. In summer the area of low pressure is over Central Asia, which produces a northerly type of wind over the islands, and this corresponds with the dry season of the year.

Take a typical cyclone moving from the Atlantic, and passing, as it often does, between Shetland and Iceland. The weather may be unnaturally warm for the season of the year, the sky becomes overcast, the wind backs to the south or south-east, and the barometer falls rapidly. Other premonitory signs may be a halo or " broch " round the sun or moon ; aurora (Merrie Dancers) high overhead ; or a heavy swell in the sea. Wind and rain follow ; after a time there may be a sudden lull ; the wind shifts to the west or north ; the barometer rises briskly and the clouds clear away, often with a strong gale of wind.

From November to February the prevailing winds are from the south and west, averaging about 5 days for each month, while October has an average of 7 days from the north. From March to June, the prevailing winds are from the north and north-east, which accounts for the cold spring and late summer, the

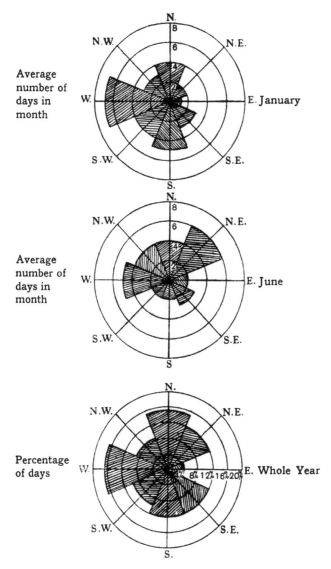

Average
number of
days in
month

N.
N.W.   N.E.
W.   E. January
S.W.   S.E.
S.

Average
number of
days in
month

N.
N.W.   N.E.
W.   E. June
S.W.   S.E.
S

Percentage
of days

N.
N.W.   N.E.
W   E. Whole Year
S.W.   S.E.
S.

Wind Roses showing the prevailing winds at
Sumburgh Head

average being about 4¾ days for each of these months. July is equally divided between west and north, having

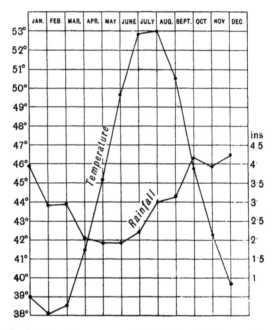

Graph showing the average Rainfall and Temperature for each month of the year, from observations taken at Sumburgh Head, extending over 35 years

5 days of each; while August has a total of 5 days from the north and September 6 from the west. The wind-rose shows that for the whole year the west wind prevails with 17 per cent, while the east has only 4.

January has 7 days of west wind and only 1 of
east ; June has 6 days from the north-east, and 4
each from north and north-west. The average number
of days in the year with calm weather is 27, with
gales 37.

As might be expected from its insular position, the
climate of Shetland is moist, and the annual rainfall
is about 37 inches, while the number of rainy days in the
year amounts to 251. This is the record for Sumburgh
Head, extending over a period of thirty-five years ;
but of course the rainfall varies considerably in different
parts of the islands. The wet period of the year embraces
the months of October, November, December and
January ; while the dry period, if it can be called
such, includes May, June and July—May being driest.
Heavy falls of snow are not uncommon ; but owing to
the sea-breezes and changeable weather, snow seldom
lies for any length of time.

The climate is somewhat cold, the mean temperature
for the year being 44.7 degrees, while the mean monthly
temperature varies from 38° in February and March,
the coldest months of the year, to 53° in August, the
warmest month. It is found that this mean annual
temperature closely approximates to the temperature
of the sea (from a depth of 100 fathoms to the bottom)
between the Hebrides and Faroe. Thunderstorms
are not frequent, and heat waves are usually followed
by fog, which is very prevalent in summer and greatly
impedes navigation round the coast. The month of
May has an average of 8 days of fog, June 12, July and

August 10 each, while the whole year shows a total of 75 days.

Winter in Shetland is dark and stormy ; but this is

Sunrise at Midsummer, 2.30 a.m.

to a certain extent compensated for by the long days of summer, from the middle of May to the end of July, when darkness is unknown. That season gives to the islands the poetic name of the " Land of the Simmer Dim."

10. People—Race, Language, Population

That Shetland was inhabited from remote times is evident from the number of primitive stone implements found all over the islands. The rude hammer and axe ; the finely-polished celts and knives ; the stone circles and brochs ; the burial mounds, with urns and stone coffins—all bear mute testimony to successive stages of progress towards civilisation. Who the earliest inhabitants were it is not easy to discover ; but it is generally agreed that for a number of centuries the Picts held sway till supplanted by invaders from Norway. Norse influence began with Viking raids, and then towards the end of the ninth century assumed the form of conquest. Scandinavian domination lasted till 1468. During this period the people were Norse to a large extent in blood and altogether in language, manners and customs ; and although Scottish influence has brought about many changes, yet the Shetlander still retains many of the Norse characteristics of his ancestors.

The spoken language is a Scottish dialect with a mixture of Norse words—many thousands, says Jacob Jacobsen—with accent and pronunication distinctly Scandinavian. Some of its peculiarities are as follows : *th* becomes *d* or *t*, as *dat eart* for *that earth* ; *qu* as in *squander* is sounded like *wh* ; long *o* is shortened ; *oo* in *good* is modified (something like French *u*) and written *ü* as *güd*. This *ü* is the shibboleth of the dialect, and is extremely difficult for any but a Shetlander to pronounce.

Nearly all the place names are of Norse origin. Islands and rocks are denoted by *uy, holm, baa, skerry, drong, stack*; openings along the coast by *voe, wick, firth, ham, hoob, min, gio, gloop, helyer*; capes by *noss, noop* or *neep, bard, mool, ness, taing, hevda* or *hevdi*, rocks or cliffs by *clett, hellya, berry, bakka, berg, ord*; inland heights by *wart* or *ward, vord, virdick, fell* or *fil, hool, sneug, kame, coll* or *kool, roni, björg*; valleys by *wall* or *vel, dal, grave* or *gref, gil, boiten, koppa, sloag, quarf, wham*; fresh water by *vatn, fors, kelda, o, ljöag, brun*; crofts and townships by *seter,* or *ster, bister, skolla, taft*; enclosures by *garth, gard, gord, girt, gairdie, krü, bü, toon, pund, hoga, hag, quhey.* An isthmus is called *aid*; a Pictish broch is *burg* or *burra. Peti* means Picts; *Finni*, Finns; *Papa* or *Papil* an Irish missionary settlement.

A century ago the population was 22,379, and reached its maximum (31,670) in 1861. After that it gradually decreased, chiefly owing to emigration, to 27,911 in 1911. The density is about fifty to the square mile.

11. Agriculture and other Industries

As an agricultural county, Shetland is comparatively poor, and that for various reasons. Although in certain districts the soil is of good quality, and produces crops that compare favourably with any Scottish county, yet much of the land may be classed as indifferent or poor. Only 4 per cent. of the whole land surface is arable. Permanent pasture might be 10 per cent. Fourteen per cent. may thus be taken as the limit of profitable agriculture. The other 86 per cent. is used for grazing. Shetland is a crofting county, but most of the holdings are too small to support a family, unless the men have some subsidiary employment, either as fishermen or sailors. Some 56 per cent. of the male population are classed as farmers or fishermen, or both combined. The result of this divided attention is seen in the backward state of cultivation, up to very recent years. Another important factor was insecurity of tenure. Before the passing of the Crofters Act in 1886 there was no security, and in many cases where a crofter improved his holding. he had to pay increased rent for his pains. But matters are now growing better. The crofter has a fair rent fixed ; his tenure is secure, so long as he conforms to certain legal requirements ; and with increased facilities for placing his produce in the market, he has every inducement to give of his labour and intelligence to agricultural work. Up-to-date methods and tools are taking the place of primitive ways and out-of-date implements ; and the Board of Agriculture is helping to educate people into better methods of agriculture and dairying. School gardening, now almost universal, is also doing much to encourage the production and use of garden vegetables.

There are 352,319 acres of land in the islands ; and of these, only 15,352 acres are under cultivation ; while 35,472 acres are laid down to permanent grass. The number of holdings in the islands is 3550, giving an average of 14.3 acres to each holding. Of these, there are 793 under 5 acres ; 2021 between 5 and 15 acres ; 563 between 15 and 30 acres ; 97 between 30 and 50 acres ; 40 between 50 and 100 acres ; 30 between 100 and 300 acres ; and 6 over 300 acres. It will thus be seen that nearly 98 per cent. of the holdings are crofts under 50 acres in extent. Of the arable land, oats is the principal product, extending to 7291 acres, or nearly

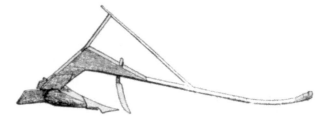

Single-Stilted Shetland Plough

one-half ; while bere is grown on 1035 acres. Potatoes take up 2795 acres and turnips the half of that, while ryegrass covers 1067 acres. There are 288,962 acres of hill pasture or " scattald " on which each crofter has the right to graze a certain number of sheep, ponies or cattle.

The native sheep are diminutive in size ; but the wool, which is made into shawls and other articles, is well known for its softness. Practically every Shetland woman is a knitter ; and although this may be reckoned a subsidiary employment, it is a very important one.

Large quantities of knitted goods are sent out of the islands every year ; and the money obtained for them is a welcome addition to the too often scanty earnings of the crofter-fishermen.   Over 2000 women are engaged in agriculture, and nearly 3000 in making hosiery.

Carding and Spinning

Besides the common grazing ground used by crofters, there are large tracts enclosed as sheep-runs, in which black-faced and other breeds are raised for the southern markets.   The total number of sheep in the islands in 1912 was 162,090.

The native breed of ponies—some of them as low as seven hands—is well known.   In days gone past they served a useful purpose as beasts of burden ; but the coming of roads and wheeled vehicles demanded a larger and stronger type of animal ; and now they are bred chiefly for export, to be used in coal mines and for other purposes. The picturesque sight of a long string of these hardy and intelligent little animals, tied head and tail,

"Leading" Home the Peats

each with " kishies " fastened to " clibbers," and the whole strapped up with a " maishie," wending their may homeward with a load of peats, is now almost a thing of the past, except in districts where roads are few. Horses, large and small, number 5827.

The native cow, like the sheep and the pony, is also diminutive ; but under favourable conditions gives a good supply of milk of excellent quality. There are 15,932 cattle in the islands, of which 6027 are milch-cows.

Except farming and fishing—with their allied occupations—industries are few in Shetland. Lerwick has two saw-mills. Freestone is quarried at the Knab near Lerwick and at the Ord in Bressay. Copper ore of good quality used to be mined at Sandlodge, and chromate of iron in Unst ; but the low prices for these made working unprofitable, and now only a small quantity of chromate is raised. Another bygone industry is the making of kelp from seaweed—once a source of considerable wealth. It may be revived, as the war (1914) caused a scarcity of potash for use in farming.

12. Fishing

For centuries it has been the custom—lately, however, to a less extent—for the crofter-fisherman to fish in the summer and to work on his farm during the rest of the year. Every voe and creek had its fleet of small boats engaged in line fishing ; but owing to the depredations of trawlers on the fishing grounds and other adverse circumstances, the system gradually declined. Line industry has now shrunk to very small dimensions, and herring fishing has taken its place. At a few creeks round the coast, small boats still engage in line fishing, while at Lerwick and Scalloway there is a considerable fleet of boats engaged in haddock and long-line fishing.

Haddocks and halibut are sent in ice to Aberdeen, while other kinds of fish are salted and dried for export. Lerwick and Scalloway have also kippering kilns.

From its position and other natural advantages, Lerwick is an ideal fishing centre, and is now one of the chief fishery ports in Scotland. The Harbour Trustees

Dutch Fishing Fleet in Lerwick Harbour

have provided pier accommodation, erected a fish market, and in other ways met the demands of this industry.

Round the harbour and along the shores of Bressay Sound a large number of curing yards are erected, each with its wooden jetty, at which the herrings are discharged. There they are taken in hand by a staff of men and women, who clean and salt them in barrels.

Steam Drifters and Fish Market, Lerwick

They used to be shipped to the Continent—chiefly to Russia and Germany—through Hamburg and the Baltic ports of Petrograd, Libau, Riga, Stettin, Konigsberg, and Dantzig.

The fish-offal is collected from the various fishing stations and taken to the factories in Bressay, where up-to-date machinery converts the raw material into articles of commerce, as fish meal for feeding cattle and pigs ; oil for tanning ; stearin for soap-making ; and manure.

Other fishing ports are Baltasound, Sandwick, Whalsay and Scalloway. There and at other centres herring fishing is carried on by sailing boats as well as steam drifters. In other districts deserted curing yards may be seen—relics of the formerly prosperous fishing stations.

The number of boats of all kinds fishing round the Shetland coast in 1913 was 952, and of these 551 were sailing boats with 2332 native men and boys. The number of drifters working from Lerwick and other ports was 380, with 3800 men, mostly Scottish and English fishermen. In addition, there were 5 drifters and 16 motor-boats owned locally, with 99 native and 10 non-resident men engaged. The total value of all these vessels, with their gear, was estimated at £1,045,839. The total quantity of fish landed amounted to 38,585 tons, valued at £347,894. This included shell-fish, herring, mackerel, ling, cod, tusk, saithe, haddock, whiting, halibut, skate, plaice, and dabs. Besides the 6241 persons actually engaged in fishing, this important industry gave employment to about 4000 other workers—gutters, coopers, carters, labourers, and sailors.

Shetland had three whaling stations, Olna Firth,

Shoal of Whales

Ronas Voe, and Colla Firth. The men employed were mostly Norwegians.

13. Shipping and Trade

Long after Shetland was annexed to Scotland, trade and friendly intercourse continued to be carried on with Norway and other countries across the North Sea. Dutch and Flemish fishermen also frequented the islands, and established a considerable trade, exchanging foreign produce for fish and articles of native manufacture. There was regular communication with Bergen, Hamburg, Bremen and other Continental ports ; and

A Busy Day at Victoria Pier, Lerwick

people from Shetland often travelled to Scotland and England by way of the Continent. As time went on, more direct communication was established with Scotland by means of sailing vessels. Trade was carried on by these till the advent of a steamship in 1836, when the paddle-boat *Sovereign* began to ply between Granton and Lerwick once a fortnight—

afterwards once a week—calling at Aberdeen, Wick and Kirkwall. At the present time four steamers regularly arrive from Leith or Aberdeen each week during summer. One of these makes a weekly trip to Scalloway and ports on the west side. Lerwick is the port of call on the east. In addition to these, another of the North of Scotland Company's steamers does the coast-wise trade on the east side and to the North Isles.

All goods, mails and passengers to and from Scotland are carried by this company's steamers. The imports consist of meal and flour; tea, sugar and butter; feeding-stuffs for cattle; and the miscellaneous articles that form the stock-in-trade of the draper, the grocer, the ironmonger, and the general merchant's store in town and country. The exports include eggs, dried and fresh fish, wool, hosiery, sheep, ponies and cattle, and cured herrings. Timber is brought direct from Norway. Coal, salt and empty barrels are imported in specially chartered vessels.

14. History

Little is known for certain of Shetland in early times. If by Thule, visited and described by Pytheas of Massilia, the ancients meant Shetland, then the first mention of the islands in our era is when Tacitus, telling of Agricola's fleet in 84 A.D. says " Dispecta est et Thule," assuming that it was really part of Mainland that the sailors descried in the dim distance. Irish missionaries christianised the natives in the sixth and seventh

K

centuries. The nearness of the islands to Norway naturally led to visits from the Viking raiders and finally to conquest and settlement by them in the ninth century. The Norsemen found two races of people—the Peti or Picts, and the Papae, descendants of the Irish missionaries. Whether these were exterminated or absorbed by the invaders is disputed; but Christianity disappeared before Odin, Thor and other Scandinavian deities.

For several centuries after this, the history of Shetland is hardly separable from the history of Orkney, and that has been already narrated.[1] Though both groups of islands formed one earldom, Orkney was the predominant partner. Shetland received scant attention except as a recruiting ground for fleets or armies. From 1195 to 1379 Shetland was disjoined from Orkney; and, while the latter came more and more under Scottish influence, the former remained under Norse. It was during the period of separation that Norse influence impressed itself most strongly on Shetland.

Norse Shetland had its Althing or Parliament, meeting on the holm in Tingwall Loch under the presidency of the Foud. Each district had its lesser Thing (as Aithsting, Nesting) and its under-Foud, who selected Raadmen or Councillors, and was assisted by the Lawmen or Legal Assessors. The Lawrightman superintended the weights and measures—an important office, since rents and duties were paid in kind. The Ranselman was a kind of parish constable.

[1] See pp. 43 *sqq.*

15. Antiquities

The most conspicuous remains of antiquity in the islands are the Pictish castles or brochs,[1] of which there

Broch of Mousa

are seventy or eighty. Most of them are in ruins ; but the one notable exception is that of Mousa Castle, the most perfect of its kind in existence. Another broch is in the loch of Clickimin. Though only a remnant, it conveys a good idea of the massive structure of these buildings.

[1] See pp. 59 *sqq.*

Standing stones occur in every parish. Other prehistoric relics are the stone circles, the earth-houses or underground dwellings, and " pechts knowes." The

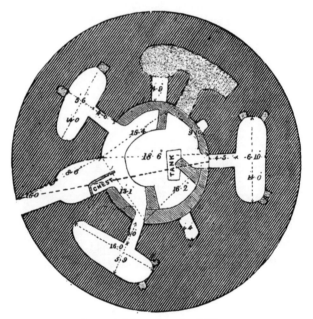

Ground Plan, Broch of Mousa

last are artificial mounds of burnt stones and earth. In some of these are found stone coffins or cists, in others urns containing the ashes of the dead.

Implements and weapons of the Stone Age are being continually unearthed. Some are rough and include hammers, clubs, whorls for spinning, stones for pounding corn, whetstones, vessels for liquids. Others are polished, and show a great advance on the rough in

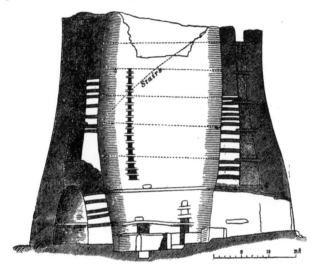

Sectional Elevation, Broch of Mousa

workmanship. These include celts or axes of porphyry or serpentine, locally known as "thunderbolts" and held in veneration by the finder. Another polished weapon is the knife, said to be found only in Shetland.

Gold, silver and bronze ornaments of the Viking Age are occasionally discovered—one of the most recent and beautiful a bracelet of gold in the isle of Oxna.

Inscribed or sculptured stones are of two kinds—Celtic, and Norse or runic. Examples of the Celtic are the Bressay Stone, the St Ninian Stone, the Lunnasting Stone, with Ogham inscriptions, and the richly sculptured monument found in Burra Isle. These

Gold Armlet (Norse) from Isle of Oxna

are all indications of the Christianity of the pre-Norse days. Four runic stones have been discovered, three in Cunningsburgh and one in Northmavine.

Bressay Stone
(Obverse)

Bressay Stone
(Reverse)

Portion of Sandstone Slab with Ogham Inscription
from Cunningsburgh

Lunnasting Stone   Burra Stone

16. Architecture

Scalloway Castle, and Muness Castle in Unst, are the only two feudal structures in the islands.

Scalloway Castle

Scalloway Castle was built by Earl Patrick Stewart in 1600, at a time when Scalloway was the capital and the only town in Shetland. Over the gateway was a Latin inscription, adapted from the New Testament contrast between the house founded on a rock and the house founded on the sand—

> " Patricius Orchadiae et Zetlandiae Comes.
> Cujus fundamen saxum est, domus illa manebit ;
> Labilis, e contra, si sit arena, perit. A.D. 1600."

Patrick's house in Sumburgh had been built on sand and had collapsed, it is said.

Muness Castle dates from 1598 and belonged to Laurence Bruce of Caltmalindie, Lord Robert Stewart's half-brother. He was at one time Grand Foud of

Muness Castle, Unst

Shetland, but for his oppressive rule he was brought to trial at Tingwall and deposed from office. As no one could conscientiously praise him, he did so himself in the following inscription over the Castle doorway :—

> " List ye to knaw this building quha began,
> Laurance the Brus he was that worthy man,
> Quha earnestlie his ayris and offspring prayis
> To help and not to hurt this wark alwayis.
>   The zeir of God 1598."

At Sumburgh is the ruin of Jarlshoff, once a residence of Lord Robert Stewart, and rendered famous in Sir Walter Scott's novel *The Pirate*.

The only building of architectural significance in

Town Hall, Lerwick

(Midnight, June)

Lerwick is the Town Hall, a handsome Gothic building, erected in 1882. Its external decorations include the arms of the town and of the nobles who were connected with the government of Shetland. The main hall is the chief centre of interest. Its beautiful stained-glass windows present a magnificent series of pictures representing the history of the islands from the time of the Norwegian conquest to James III of Scotland, who married Margaret of Denmark.

17. Communications

Till the middle of last century Shetland was almost devoid of roads. All traffic had to go by water, while travelling by land was on foot or on horseback over moorland tracks. The failure of the potato crop in 1846 and following years caused much distress in the islands ; and the food and money sent by the Board for the Relief of Destitution in the Highlands enabled labour to be hired for road-making. Between 1849 and 1852 about 120 miles of roads were constructed, joining Lerwick with Dunrossness on the south, with Scalloway and Walls on the west, and with Lunna, Mossbank and Hillswick on the north ; while a road 17 miles long was constructed through Yell. Under the Zetland Roads Act (1864) roads were improved and extended all over the islands. The County Council now manages all roads and bridges.

The southern terminus of the main road is Grutness, near Sumburgh. From this point to Lerwick the road skirts the east side of the island, with branches leading to various districts. From Lerwick the north road begins by climbing the hill of Fitch. At the bridge of Fitch a branch diverges to Scalloway ; and further on, at Tingwall, is the junction of another route to the ancient capital. Here the main road bifurcates, one fork going in a westerly direction through Whiteness,

Weisdale, Aithsting, Walls and Sandness, and terminating at Huxter. The other fork after passing Girlsta and Sandwater, enters the dreary Lang Kame, a stretch of five miles without a single habitation, where, says superstition, the benighted traveller may meet " Da Trows," the goblins of Norse mythology. The road then runs by the Loch of Voe to Olnafirth Kirk. Here it divides—going on the right by Dales Voe to Mossbank, on the left to the narrow neck between Busta Voe and Sullom Voe and thence by Mavis Grind past Punds Water. Again it branches, to Ollaberry on the right and Hillswick on the left. In Yell the chief road runs from Burra Voe to Cullivoe in the north-east. The trunk road in Unst—the best in the islands—stretches from Uyeasound to Norwick.

18. Roll of Honour

Although Shetland has produced no names of world-wide celebrity, yet many sons of the " Old Rock " have risen to distinction both at home and abroad.

The first place must be given to Arthur Anderson, (1792–1868), who, commencing life as a humble fish-worker at Gremista, near Lerwick, was in 1840 one of the founders of the P. and O. Steamship Company, and ultimately its chairman. Much of his wealth he spent on Shetland. He established the first newspaper in the islands—*The Shetland Journal* ; he started the Shetland Fishing Company, which largely helped to emancipate the crofter-fishermen from their bondage to the landlords ; he introduced Shetland hosiery to the outside world ; he was influential in securing steam communication between the islands and Scotland ; and,

Arthur Anderson

not to mention more of his benefactions, he founded the Anderson Educational Institute, which is now administered by the Education Authority as a Higher Grade School and Junior Student Centre for the county. Another benefactor was R. P. Gilbertson, a colonial merchant, who presented the Gilbertston Park to Lerwick and founded the Gilbertson Trust for the benefit of Shetlanders.

The islands boast a goodly array of writers, and among these the Edmonston family stands conspicuous. Dr Arthur Edmonston (1775–1841) wrote *A View of the*

Anderson Institute, Lerwick

Ancient and Present State of the Zetland Islands ; his brother Laurence (1795–1879) was a distinguished Scandinavian scholar and the author of many papers on natural history ; Laurence's son, Thomas, born in 1825, was the discoverer of the Shetland plant *arenaria Norvegica,* and the author of a *Flora of Shetland.* At the age of twenty he was elected Professor of Botany and Natural History in Anderson's College at Glasgow ; but next year he was accidentally shot dead in Peru, while on a scientific expedition to the Pacific ; Thomas's brother, Biot (1827–1906) was joint-author with his sister Mrs Saxby, of *The Home of a Naturalist*. Dr Saxby wrote *The Birds of Shetland*. Thomas Gifford of Busta, who died in 1760, was the author of *Historical Description of the Zetland Isles ;* Dr Robert Cowie, of *Shetland : Descriptive and Historical ;* Dr Copeland, of *A Dictionary of Medicine ;* Miss Spence, of *Earl Rognvald and his Forebears*, and a *Memoir of Arthur Laurenson*, a scholar deeply versed in Scandinavian lore ; Gilbert Goudie, of *Antiquities of Shetland*, and other works ; and John Spence, of *Shetland Folk Lore*. Of minor poets and vernacular writers we may name Basil R. Anderson ; Laurence J. Nicolson, " The Bard of Thule " ; George Stewart ; T. P. Ollason ; and J. B. Laurence.

Of men who have risen to distinction in the Government service, the best known is Sir Robert G. C. Hamilton (1836–1895), a son of Dr Hamilton of Bressay. He was at various times Accountant-General of the Navy, Under-Secretary for Ireland, Governor of Tasmania, and Chairman of the Board of Customs.

19. The Chief Towns and Villages of Shetland

(The figures in brackets after each name give the population in 1911, and those at the end of each section are references to pages in the text.)

Balta Sound is a fishing port on the east of Unst. It was formerly a prosperous centre for herrings. (pp. 123, 142.)

Hillswick is a seaport in Northmavine parish in the northwest of Mainland. Hillswick Voe affords sheltered anchorage for vessels. (pp. 125, 156, 157.)

Lerwick (4664), the capital and the only burgh of the county, and the most northerly town in Britain, lies on the west side of Bressay Sound, which forms a safe and commodious harbour. During the Dutch War, Cromwell built and garrisoned the fort. This may be taken as the beginning of the town. In 1781 the fort was put into a state of defence, and named Fort Charlotte in honour of George III's Queen. It is now a Coast Guard Station and Royal Naval Reserve Headquarters. The Old Town, built on the side of a hill facing the harbour, has one narrow and irregular street running parallel to the shore, with numerous lanes branching off at right angles. Some of the houses at the south end of the town are built right into the sea ; but elsewhere, the shore has been reclaimed, and an esplanade and wharves take the place of the " Lodberries " of former times. Victoria Pier (erected in 1888), Alexandra Wharf with the Fish Market, and the Boat Harbour, are among the latest harbour improvements. The New Town, which lies on the landward side of the hill, has wide streets and modern houses, the Central School being the principal building ; while near by is the Gilbert Bain Hospital. The Docks are situated at the north end of the town, and here also are boat-building sheds, saw-mills and barrel factories. Adjoining the

L

Lerwick, N.

Lerwick, S.

Anderson Institute is a Hostel for country girl-bursars attending the Institute. This handsome building, the gift of Robert H. Bruce, Esq., of Sumburgh, is the first building of its kind in Scotland. Lerwick has two weekly

Scalloway.

newspapers—*The Shetland Times* and *The Shetland News*. At Sound, in the vicinity, is a Government wireless station. (pp. 103, 118, 120, 139, 140-4-5, 155-6-7-9.)

Scalloway (824), an ancient village, was at one time the capital. It is the chief port of call for steamers on the west, and is a centre for herring and white fishing. The Castle is the chief object of interest. (pp. 108, 110, 126, 139, 140-2-5, 153, 156.)

Sandwick, in Dunrossness, is a busy fishing centre. (pp. 117, 142.)

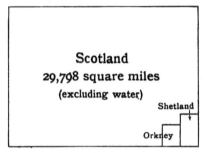

Fig. 1.   Area of Orkney and of Shetland
compared with that of Scotland

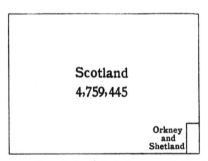

Fig. 2.   Population of Orkney and Shetland
compared with that of Scotland

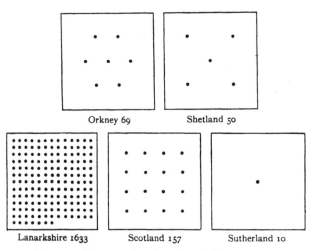

Orkney 69   Shetland 50

Lanarkshire 1633   Scotland 157   Sutherland 10

Fig. 3. Comparative density of Population
to the square mile

(Each dot represents 10 persons)

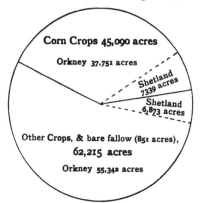

Corn Crops 45,090 acres

Orkney 37,751 acres

Shetland 7339 acres

Shetland 6,873 acres

Other Crops, & bare fallow (851 acres),
62,215 acres

Orkney 55,342 acres

Fig. 4. Proportionate areas of Corn and
other Cultivations

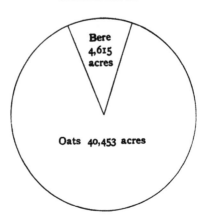

Fig. 5.   Proportionate areas of
Chief Cereals

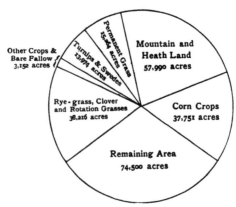

Fig. 6.   Proportionate areas of land—Orkney

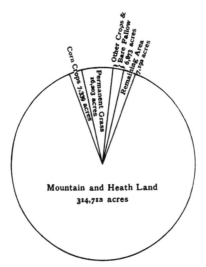

Corn Crops 7,326 acres
Permanent Grass 16,903 acres
Other Crops & Bare Fallow 6,973 acres
Remaining Area 7,192 acres

Mountain and Heath Land
314,712 acres

Fig 7. Proportionate areas of land—Shetland

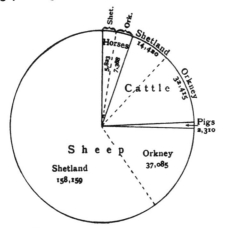

Shet. Ork.
Horses
Shetland 14,420
5,823 7,966
Orkney 32,415
Cattle
Pigs 2,310
Sheep Orkney 37,085
Shetland 158,159

Fig. 8. Proportionate numbers of Live Stock

Milton Keynes UK
Ingram Content Group UK Ltd.
UKHW032321161024
449665UK00001B/12